Generis

PUBLISHING

Octav Olteanu

Recent Results on Markov Moment Problem, Polynomial Approximation and Related Fields in Analysis

CIP a Camerei Naționale a Cărții

Olteanu, Octav.

Recent Results on Markov Moment Problem, Polynomial : Approximation and Related Fields in Analysis / Octav Olteanu. – Chișinău : Generis Publishing, 2020 (Print on demand). – 94 p.

Referințe bibliogr.: p. 89-93 (67 tit.).

ISBN 978-9975-153-22-5.

519.217

O-50

Cover image: www.pixabay.com

Generis Publishing
Online orders: www.generis-publishing.com
Orders by email: info@generis-publishing.com

Preface

The main subjects of the present book are: applications of Hahn-Banach type extension theorems for linear operators preserving two constraints and of polynomial approximation on unbounded subsets to the existence and uniqueness of the solutions for some Markov moment problems; convex optimization for functions and operators; a global version of Newton's method for convex functions and operators; invariant subspaces and balls of bounded linear operators; proving results from linear analysis for convex operators; topological versions for sandwich results on finite simplicial sets, a direct proof for a generalization of Hahn-Banach theorem and concluding remarks. Most of our results involve concrete spaces and operators. A special attention is accorded to the relationship between a global version of Newton's method for convex operators and the contraction principle. The last two chapters contain unpublished new research. The other chapters are partially based on recently published results contained in the references [40]-[55]. Most of earlier results are recalled without repeating their proofs. However, some of the basic results such as those from [3] and [35] are proved in detail. Mazur-Orlicz theorem in concrete spaces is also under attention.

July 10, 2020

The Author

Contents

INTRODUCTION

The aim of the present book is to emphasize recent results in the following six actual research fields, presented in the six chapters of the book; the main topics are: 1) Markov moment problem and related problems; 2) Pointing out properties, evaluating and optimizing convex functions and convex operators; 3) Proving a global Newton like methos for convex functions and operators and pointing out its connection to contraction principle; 4) Constructing invariant subspaces for a large class of bounded linear operators and related results; 5) Proving results from analysis of linear operators, for convex operators; emphasizing relationship between linear and sublinear continuous operators; extending inequalities via Krein-Milman theorem; 6) Proving topological versions of sandwich theorems of type $f \leq h \leq g$, where $f, -g$ are convex and h is affine, on bounded and unbounded special convex subsets; giving a direct sharp proof for a main generalization of Hahn-Banach theorem, accompanied by its motivations; stating and justifying cuncluding remarks. All our theorems are accompanied by examples, solving concrete problems related to basic spaces of functions and respectively concrete equations. The interested reader can find detailed proofs of recent results, as well as of our earlier basic results which were published in non-Open Access Journals. A common point of the first three chapters is the notion of convex function (or operator). In this respect, the linear solution of a Markov moment problem is dominated by a convex operator and is minorated by the null operator on the convex cone of the domain space. This last property is called positivity of the solution. The dominating convex continuous operator controls the norm of the linear solution. Recall that a Markov moment problem is an interpolation problem with two constraints on the linear solution. For details, see the Introduction of the first chapter. The existence of such a solution is proved by means of Hahn-Banach type theorems and their generalizations. In this respect, we use the main results first published in [34], [36], [37]. All necessary such earlier results are recalled without their proofs. A direct sharp proof of a basic such result and its recent motivations is given in Chapter 6. This last chapter is directly connected to the first chapter. Another important aspect of the Markov moment problem is the uniqueness of the solution. Sometimes the uniqueness follows from the proof of the exsitence for the solution, via polynomial approximation results proved in [27], [42], [44], [50], [52], [54] and in other works. Since the values of the linear continuous solution on polynomials are prescribed, the uniqueness of the solution follows whenever the polynomials are dense in the domain function space. The most interesting case is that of of function spaces in several real variables, for which polynomial approximation on arbitrary (unbounded) closed subsets have been completely proved firstly in [42] and completed in [44], [50]. The method works for any Cartesian product of closed intervals. In particular, it works for spaces such as $L_\mu^1(\mathbb{R}^n), L_\mu^1([0, \infty)^n)$, etc., where μ is a positive Borel $M -$ determinate (moment determinate) measure. Recall that a measure is called $M -$ determinate if it is uniquely determinate by its classical moments, or, equivalently, by its values on polynomials. Our polynomial approximation results partially solve the difficulty arising from the fact that there exist nonnegative polynomials on $\mathbb{R}^n, n \geq 2$, which are not expressible as sums of squares. This way, some of the conditions in our statements are formulated in terms of quadratic forms. On the other hand, these approximation methods lead to a characterization of positivity of some bounded linear operators only in terms of quadratic forms. Some other applications of polynomial approximation appear in Chapter 4, where they are not related to the moment problem. Going back to Chapter 1, it is worth noticing that a special attention is focused on Mazur-Orlicz theorem in concrete spaces. Chapter 2 is devoted to evaluating and optimizing convex functions and operators, under certain convex or even linear constraints. The results are presented in chronological order ([35], [49], [51], [54]). For Pareto optimization and related applications see [8], [9]. For further results on convex functions and operators see [21], [30], [31]. Chapter 3 deals with Newton's method for convex operators. The Newton iterations are well-known. The problem is that the obtained sequence is not always convergent. In fact, Newton's method works sometimes only locally. When the involved function (or operator) is convex and isotone (or anti-isotone), with continuous derivative of first order, the method leads to a rapidly convergent sequence. Its limit is the zero of the given function (respectively operator), which can be approximated with the control of the norm of the error. The strength of this method consists in the fact that is global, and the weakness is its limitation to convex (or concave) non-decreasing or non-increasing operators. The connection with the contraction principle is emphasized and proved in detail. In some cases, approximation in terms of a contraction constant is preferable to that from the classical Newton method (see [3], [54], [55]). Chapter 4 is devoted to construction of invariant (closed) subspaces for a class of bounded linear operators. A related differential equation is studied and discussed in this respect. Invariance of the unit ball of some L^1 spaces is also under attention. Here polynomial approximation is applied once more. It allows passing from inequalities verified on special nonnegative polynomials, to the same inequalities on arbitrary nonnegative functions in the domain space. Chapters 5 and 6 (the last ones) contain unpublished results. In Chapter 5, a uniformly boundedness property for classes of convex operators is proved and its relationship with previous related results in the literature is briefly discussed. As an important particular case, we point out applications of this result to classes of sublinear operators. Extending inequalities via Krein-Milman theorem is considered in the end of this chapter. The first aim of Chapter 6 is to prove topological versions for sandwich results $f \leq h \leq g$, where $f, -g$ are convex and h is affine on X. The case when X is a Choquet simplex, as well as that of X being a finite simplicial set [2] is under attention. The main results on this subject are Theorems 2.3 and 2.1. Secondly, applications of Krein-Milman and Carathéodory's theorem are emphasized. In the end, a direct sharp proof of a generalization of Hahn-Banach theorem is detailed and motivated. Useful comments and remarks conclude the book.

We hope that this work could be useful for researchers, PhD students and Post Doc researchers, students in the last two years of Mathematics Faculties in Universities, professors, engineers and any reader interested in mathematical analysis and its applications.

July 10, 2020

The Author

CHAPTER 1

ON MARKOV MOMENT PROBLEM AND MAZUR-ORLICZ THEOREM

1. Introduction

We recall the classical formulation of the moment problem, under the terms of T. Stieltjes, given in 1894-1895 (see the basic book of N.I. Akhiezer [1] for details): find the repartition of the positive mass on the nonnegative semi-axis, if the moments of arbitrary orders k ($k = 0,1,2, ...$) are given. Precisely, in the Stieltjes moment problem, a sequence of real numbers $(s_k)_{k\geq0}$ is given and one looks for a nondecreasing real function $\sigma(t)$ ($t \geq 0$), which verifies the moment conditions:

$$\int_0^\infty t^k \, d\sigma = s_k, \qquad (k = 0,1,2, ...)$$

This is a one dimensional moment problem, on an unbounded interval. Namely, is an interpolation problem with the constraint on the positivity of the measure $d\sigma$. The numbers $s_k, k \in \mathbb{N} = \{0,1,2, ...\}$ are called the moments of the measure $d\sigma$. Existence, uniqueness and construction of the solution σ are studied. The present work concerns firstly the existence problem. The connection with the positive polynomials and extensions of linear positive functional and operators is quite clear. Namely, if one denotes by $\varphi_j, \varphi_j(t) := t^j, j \in \mathbb{N}, t \in [0, \infty)$, $\mathcal{P}$ the vector space of polynomials with real coefficients and

$$T_0: \mathcal{P} \to \mathbb{R}, T_0\left(\sum_{j \in J_0} \alpha_j \varphi_j\right) := \sum_{j \in J_0} \alpha_j s_j,$$

where $J_0 \subset \mathbb{N}$ is a finite subset, then the moment conditions $T_0(\varphi_j) = s_j, j \in \mathbb{N}$ are clearly verified. It remains to check whether the linear form T_0 has nonnegative values at nonnegative polynomials. If the latter condition is also accomplished, then one looks for the existence of a linear positive extension T of T_0 to a larger ordered function space X which contains both $\mathcal{P}$ and the space of continuous compactly supported functions, then representing T by means of a positive regular Borel measure μ on $[0, \infty)$, via Riesz representation theorem. Alternately one can apply directly Haviland theorem [19]. If an interval (for example $[a, b]$, $\mathbb{R}$, or $[0, \infty)$) is replaced by a closed subset of $\mathbb{R}^n, n \geq 2$, we have a multidimensional moment problem. Passing to an example of the multidimensional real classical moment problem, let denote

3

$$\varphi_j(t) = t^j = t_1^{j_1} \cdots t_n^{j_n}, \qquad j = (j_1, \ldots, j_n) \in \mathbb{N}^n, t = (t_1, \ldots, t_n) \in \mathbb{R}_+^n, n \in \mathbb{N}, n \geq 2 \quad (1,1)$$

If a sequence $\left(y_j\right)_{j \in \mathbb{N}^n}$ is given, one studies the existence, uniqueness and construction of a linear positive form F defined on a function spaces containing polynomials, such that the moment conditions

$$F\left(\varphi_j\right) = y_j, \qquad j \in \mathbb{N}^n \tag{1.2}$$

be accomplished. Usually, the positive linear form F can be represented by means of a positive regular Borel measure on $\mathbb{R}_+^n$. when un upper constraint on the solution F is required too, we have a Markov moment problem. This requirement is formulated as F being dominated by a convex functional, which might be a norm, and its aim is to control the continuity and the norm of the solution. All these aspects motivate the study sketched in the next section, which is mainly devoted to the abstract moment problem. Clearly, the classical (Stieltjes) moment problem is an extension problem for linear functionals, from the subspace of polynomials to a function space which contains both polynomials as well as the continuous compactly supported real functions on $\mathbb{R}_+^n$. From solutions linear functionals, many authors considered solutions linear operators. Of course, in this case the moments $y_j, j \in \mathbb{N}^n$ are elements of an ordered vector space Y (usually Y is an order complete Banach lattice). The order completeness is necessary in order to apply Hahn-Banach type results for operators defined on polynomials and having Y as codomain. Various aspects of the classical moment problem have been studied (see the References). The case of multidimensional moment problem on compact semi-algebraic subsets in $\mathbb{R}^n$ was intensively studied. Clearly, the classical moment problem is related to the form of positive polynomials on the involved closed subset of $\mathbb{R}^n$. As it is well-known, there exists nonnegative polynomials on the entire space $\mathbb{R}^n, n \geq 2$, which are not sums of squares of polynomials, contrary to the case $n = 1$. The analytic form of positive polynomials on special closed unbounded finite dimensional subsets is crucial in solving classical moment problems on such subsets (see [25] for the expression of nonnegative polynomials on a strip, in terms of sums of squares). Such results are useful in characterizing the existence of a positive solution by means of signatures of quadratic forms. In case of Markov moment problem, approximation of nonnegative compactly supported continuous functions (with their support contained in a closed subset) by special nonnegative polynomials on that subset, having known their analytic form is very important. A main such approximation result over closed unbounded subsets of $\mathbb{R}^n, n \geq 1$ was first proved in [42], Lemma 7, and completed in [44], [50], [52], [54]. Based on the form of nonnegative polynomials on unbounded intervals and the above mentioned approximation results, Markov moment problem on Cartesian products of unbounded intervals can be solved in terms of products of quadratic forms. In most of the cases, the uniqueness of the solution of the Markov moment problem on spaces $L_\nu^1(A)$ follows too, thanks to the density of polynomials in such spaces; here A is a closed unbounded subset of $\mathbb{R}^n$ and μ is a positive regular M-determinate Borel measure on A. Recall that a measure is $M-$determinate (moment determinate), if it is uniquely determinate by its classical moments (or, equivalently, by its values on polynomials). For determinacy, uniqueness and non-uniqueness of solutions of some moment problems see [15], [19], [61], [62]. For the construction of some solutions see [16], [26], [32], [40], [47], [50]. Connections of the moment problem to operator theory are emphasized in [15], [22], [57], [61], [64]. Interesting connections of some moment problem to fixed point theory are pointed out in [5] and [6]. To conclude, for characterizing the existence of a solution for a classical moment problem in terms of moments, extension Hahn-Banach results and their generalizations accompanied by knowing the analytic form of positive polynomial on the set under discussion are the basic tools. Sometimes, especially in Markov moment problem, the uniqueness of the (continuous linear) solution follows too, by the proof of its existence, thanks to the density of polynomials in some function spaces (even over an unbounded closed subset). In this respect, for both existence and uniqueness of the solution, polynomial approximation on unbounded subsets is essential. Otherwise, the uniqueness problem requires specific methods. Basic monographs on the moment problem are [1], [20], [61]. Connections of Markov moment problem with optimization are studied in [32], [49] (see also the references therein). The monographs [13], [29], [30], [59] contain main information on Banach spaces, ordered Banach spaces, ordered topological vector spaces, convexity in finite and infinite dimensional spaces and related inequalities. In [31], the notion of a Markov operator is recalled and applied. It appears in the present chapter too, but in another context (see [53], [54]). The rest of this chapter is organized as follows. Section 2 is devoted to recalling known results and methods on vector-valued Markov moment problem and Mazur-Orlicz theorem, which are applied in the sequel. In Section 3, moment problems and Mazur-Orlicz theorems in concrete spaces are solved. Some of the solutions are Markov operators. Generally, the norms of the solutions can be determined by means of the norms of the bounded sublinear upper constraints. Section 4 is devoted to polynomial approximation on unbounded subsets and its applications to the Markov moment problem.

2. Extension of linear operators, the abstract moment problems and Mazur-Orlicz theorem

Almost all the results of this section will be applied in the sequel. Three of them have been published first in [37], using earlier results from [34], [36]. Their proofs are based on previous theorems on constrained extension of linear operators published in [34], [36]. The next result was published first in [34], where its proof was sketched as well. The detailed proof and consequences can be found in [36]. In the following statement, E will be a real vector space, F an order-complete vector lattice, $A, B \subset E$ convex subsets, $Q: A \to F$ a concave operator, $P: B \to F$ a convex operator, $H \subset E$ a vector subspace, $T_0: H \to F$ a linear operator. All vector spaces and linear operators are considered over the real field.

Theorem 2.1. ([34]). *Assume that*

$$T_0(x) \geq Q(x) \; \forall x \in H \cap A, \, T_0(x) \leq P(x) \; \forall x \in H \cap B$$

The following statements are equivalent:
(a) *there exists a linear extension* $T : E \to F$ *of the operator* T_0 *such that*
$$T|_A \geq Q, T|_B \leq P;$$

(b)*there exists* $P_1 : A \to F$ *convex and* $Q_1 : B \to F$ *concave operator such that for all*

$$(\rho, t, \lambda', a_1, a', b_1, b', v) \in [0,1]^2 \times (0, \infty) \times A^2 \times B^2 \times H ,$$

one has

$$(1-t)a_1 - tb_1 = v + \lambda'[(1-\rho)a' - \rho b'] \Rightarrow$$
$$(1-t)P_1(a_1) - tQ_1(b_1) \geq T_0(v) + \lambda'[(1-\rho)Q(a') - \rho P(b')] \tag{2.1}$$

Thus in the last relation we have a convex operator on the left hand side, and a concave operator on the right hand side.

Theorem 2.2. ([37]). *Let E be a preordered vector space, F an order complete vector lattice, $P: E \to F$ a convex operator, $\{x_j\}_{j \in J} \subset E$, $\{y_j\}_{j \in J} \subset F$ given families. The following statements are equivalent*

(a) *there exists a linear positive operator $T : E \to F$ such that*

$$T(x_j) = y_j \ \forall j \in J, \ \ T(x) \leq P(x) \ \forall x \in E ;$$

(b) *for any finite subset $J_0 \subset J$ and any $\{\lambda_j\}_{j \in J_0} \subset R$, we have*

$$\sum_{j \in J_0} \lambda_j x_j \leq x \in E \Rightarrow \sum_{j \in J_0} \lambda_j y_j \leq P(x)$$

If in addition we assume that P is isotone $\big(u \leq v \Rightarrow P(u) \leq P(v)\big)$, the assertions (a) and (b) are equivalent to (c), where

(c) *for any finite subset $J_0 \subset J$ and any $\{\lambda_j\}_{j \in J_0} \subset R$, the following inequality holds*

$$\sum_{j \in J_0} \lambda_j y_j \leq P\left(\sum_{j \in J_0} \lambda_j x_j\right)$$

Theorem 2.3. ([37]). *Let $E, F, \{x_j\}_{j \in J}, \{y_j\}_{j \in J}$ be as in Theorem 2.2, $T_1, T_2 \in L(E, F)$ two linear operators. The following statements are equivalent*

(a) *there is a linear operator $T \in L(E, F)$ such that*

$$T_1(x) \leq T(x) \leq T_2(x) \ \forall x \in E_+, T(x_j) = y_j \ \forall j \in J ;$$

(b) *for any finite subset $J_0 \subset J$ and any $\{\lambda_j\}_{j \in J_0} \subset R$, the following implication holds true*

$$\left(\sum_{j \in J_0} \lambda_j x_j = \psi_2 - \psi_1, \psi_1, \psi_2 \in E_+\right) \Rightarrow \sum_{j \in J_0} \lambda_j y_j \leq T_2(\psi_2) - T_1(\psi_1).$$

The next result is a variant of Mazur-Orlicz Theorem, where the interpolation conditions $T(x_j) = y_j \ \forall j \in J$ from Theorem 2.2 are replaced by the weaker requirements $T(x_j) \geq y_j \ \forall j \in J$.

Consequently, the corresponding weaker conditions appear in (b), (c) (see below), where it is sufficient that the coefficients $\lambda_j, j \in J_0$ to be nonnegative.

Theorem 2.4. ([37]). *Let E be a preordered vector space, F an order complete vector space $\{x_j\}_{j \in J}$, $\{y_j\}_{j \in J}$ be as in Theorem 2.2, $P: E \to F$ a sublinear operator. The following statements are equivalent*

(a) *there exists a linear positive operator $T: E \to F$ such that*

$$T(x_j) \geq y_j \ \forall j \in J, \ T(x) \leq P(x) \ \forall x \in E;$$

(b) *for any finite subset $J_0 \subset J$ and any $\{\lambda_j\}_{j \in J_0} \subset R_+ = [0, \infty)$, the following implication holds true*

$$\sum_{j \in J_0} \lambda_j x_j \leq x \in E \Rightarrow \sum_{j \in J_0} \lambda_j y_j \leq P(x)$$

If in addition we assume that P is isotone, the assertions (a) and (b) are equivalent to (c), where

(c) *for any finite subset $J_0 \subset J$ and any $\{\lambda_j\}_{j \in J_0} \subset R_+$, the following inequality holds*

$$\sum_{j \in J_0} \lambda_j y_j \leq P\left(\sum_{j \in J_0} \lambda_j x_j\right)$$

The next theorem of this section is an earlier extension result, sometimes called Lemma of the majorizing subspace, for positive linear operators on subspaces in ordered vector spaces (X, X_+), for which the positive cone X_+ is generating ($X = X_+ - X_+$). Recall that in such an ordered vector space X, a vector subspace S is called a majorizing subspace if for any $x \in X$, there exists $s \in S$ such that $x \leq s$.

Theorem 2.5. ([13], [59], [61]). *Let X be an ordered vector space whose positive cone is generating, $S \subset X$ a majorizing vector subspace, Y an order complete vector lattice, $F_0: S \to Y$ a linear positive operator. Then F_0 has a linear positive extension $F: X \to Y$ at least.*

Our next goal is to find a variant of Theorem 2.3, when the linear operators involved in the constraints on the solution T are more general. Namely, T_2 will be replaced by an arbitrary sublinear operator, T_1 will be a supralinear operator, and the positive cone E_+ will be an arbitrary convex cone C (which might be equal to the whole domain space E). Consequently, the linear subspace $C \cap (-C)$ is not generally reduced to the origin.

Theorem 2.6. *Let E be a vector space, F an order complete vector lattice, $C \subset E$ an arbitrary convex cone, $\Phi: C \to F$ a sublinear operator, $Q: C \to F$ a supralinear operator. Let $H \subset E$ be a vector subspace and $T_0: H \to F$ a linear operator. Assume that $Q|_{H \cap C} \leq T_0|_{H \cap C} \leq \Phi|_{H \cap C}$. The following statements are equivalent*

(a) *there exists a linear extension $T: E \to F$ of T_0 such that $Q \leq T|_C \leq \Phi$;*

(b) *for all $(f, h_1, h_1', h_2, h_2') \in H \times C^4$, the following implication holds true*

$$f = h_1 + h_1' - (h_2 + h_2') \Rightarrow T_0(f) \le \Phi(h_1) + \Phi(h_1') - \left(Q(h_2) + Q(h_2')\right) \tag{2.2}$$

(c) *for all* $\left(f, \tilde{h}_1, \tilde{h}_2\right) \in H \times C^2$, *one has*

$$f = \tilde{h}_1 - \tilde{h}_2 \Rightarrow T_0(f) \le \Phi(\tilde{h}_1) - Q(\tilde{h}_2) \tag{2.3}$$

Proof. The equivalence $(a) \Leftrightarrow (b)$ follows directly from Theorem 2.1, applied to $A = B = C, P = P_1 := \Phi, Q_1 := Q, v = f$. The implication (2.1) can be written as

$$f = (1-t)a_1 + \lambda'\rho b' - (tb_1 + \lambda'(1-\rho)b') \Rightarrow$$

$$T_0(f) \le (1-t)\Phi(a_1) + \lambda'\rho\Phi(b') - \left(tQ(b_1) + \lambda'(1-\rho)Q(a')\right) =$$

$$\Phi\left((1-t)a_1\right) + \Phi(\lambda'\rho b') - \left(Q(tb_1) + Q(\lambda'(1-\rho)a')\right) =$$

$$\Phi(h_1) + \Phi(h_1') - \left(Q(h_2) + Q(h_2')\right),$$

where $h_1 := (1-t)a_1 \in C, h_1' := \lambda'\rho b' \in C, h_2 := tb_1 \in C, h_2' := \lambda'(1-\rho)a' \in C$. In other words, for Φ sublinear and Q supralinear, (2.1) is equivalent to (2.2). According to Theorem 2.1, (2.1) (hence (2.2)) is equivalent to the existence of a linear extension T of T_0, from E to F, with the properties stated at point (a) of the present theorem. Namely, one has $Q \le T|_C \le \Phi$. Thus $(a) \Leftrightarrow (b)$ of the present theorem is proved. To prove $(a) \Leftrightarrow (c)$, observe that $(a) \Rightarrow (c)$ is obvious, thanks to the properties of T:

$$f = \tilde{h}_1 - \tilde{h}_2, \tilde{h}_1, \tilde{h}_2 \in C \Rightarrow T_0(f) = T(f) = T(\tilde{h}_1) - T(\tilde{h}_2) \le \Phi(\tilde{h}_1) - Q(\tilde{h}_2)$$

For the implication $(c) \Rightarrow (a)$, it is sufficient to prove that $(c) \Rightarrow (b)$, which is obvious. Indeed, under the hypothesis and using the notations from (b), we have

$$f = \tilde{h}_1 - \tilde{h}_2,$$

where $\tilde{h}_1 := h_1 + h_1', \tilde{h}_2 := h_2 + h_2'$. The hypothesis (2.3) from (c) leads to

$$T_0(f) \le \Phi(h_1 + h_1') - Q(h_2 + h_2') \le \Phi(h_1) + \Phi(h_1') - \left(Q(h_2) + Q(h_2')\right)$$

Thus (b) is verified and (a) follows according to implication previously proved. This concludes the proof. $\square$

Remark 2.1. The basic implication of Theorem 2.6 is $(c) \Rightarrow (a)$.

3. Moment problems and Mazur-Orlicz theorem in concrete spaces

The aim of this Section is to point out some cases when the above results can be applied. A special care is accorded to the control of the norm of the solutions.

Theorem 3.1. *Let E be a normed vector lattice, F an order complete normed vector lattice, $\Phi: E \to F$ a bounded sublinear operator, $\{e_j; j \in J\}$ an arbitrary family of linearly independent elements of $E, \{y_j; j \in J\}$ a family of given elements of F. The following statements are equivalent*
    (a) *there exists a bounded linear operator T from E to F such that*

$$T(e_j) = y_j, j \in J, \qquad |T(h)| \leq \Phi(h) \; \forall h \in E_+, \|T\| \leq 2\|\Phi\|;$$

    (b) *for any finite subset $J_0 \subset J$ and any $\{\alpha_j; j \in J_0\} \subset \mathbb{R}, h_1, h_2 \in E_+,$ the following implication holds*

$$\sum_{j \in J_0} \alpha_j e_j = h_1 - h_2 \Rightarrow \sum_{j \in J_0} \alpha_j y_j \leq \Phi(h_1) + \Phi(h_2)$$

Proof. Condition (b) of the present statement, also using the notations of Theorem 2.6, where $H := Span\{e_j; j \in J\}$, $T_0: H \to F, T_0(\sum_{j \in J_0} \alpha_j e_j) := \sum_{j \in J_0} \alpha_j y_j, Q(h) := -\Phi(h), h \in E_+ := C$ lead to the conclusion that condition (c) of Theorem 2.6 is accomplished. Application of (c)$\Rightarrow$(a) of the latter theorem, yields the existence of a linear extension T of T_0 with the following properties

$$T(e_j) = T_0(e_j) := y_j, j \in J, -\Phi(h) \leq T(h) \leq \Phi(h), \forall h \in E_+ \Rightarrow$$

$$|T(h)| \leq \Phi(h) \; \forall h \in E_+$$

$$\Rightarrow \|T(h)\| \leq \|\Phi(h)\| \leq \|\Phi\|\|h\| \; \forall h \in E_+$$

If $x \in E$, then
$$\|T(x)\| = \|T(x^+) - T(x^-)\| \leq \|T(x^+)\| + \|T(x^-)\| \leq$$

$$\|\Phi\|(\|x^+\| + \|x^-\|) \leq 2\|\Phi\|\|x\| = 2\|\Phi\|\|x\|$$

We have obtained $\|T\| \leq 2\|\Phi\| < \infty$. In particular, T is bounded. On the other side, observe that the implication (a)$\Rightarrow$(b) is obvious. This concludes the proof. $\quad\square$

Theorem 3.2. *Under assumptions and using the notations of Theorem 3.1, additionally assume that Φ is isotone. The following statements are equivalent*

    (a) *there exists a positive linear operator $T: E \to F$ such that*

9

$$T(e_j) = y_j, j \in J, \qquad T(h) \leq \Phi(h) \ \forall h \in E, \|T\| \leq \|\Phi\|$$

(b) *for any finite subset $J_0 \subset J$ and any $\{\alpha_j; j \in J_0\} \subset \mathbb{R}$, the following inequality holds*

$$\sum_{j \in J_0} \alpha_j y_j \leq \Phi\left(\sum_{j \in J_0} \alpha_j e_j\right)$$

Proof. The implication (a)$\Rightarrow$(b) is obvious, because of the following relations:

$$\sum_{j \in J_0} \alpha_j y_j = \sum_{j \in J_0} \alpha_j T(e_j) = T\left(\sum_{j \in J_0} \alpha_j e_j\right) \leq \Phi\left(\sum_{j \in J_0} \alpha_j e_j\right)$$

To prove the converse, we apply Theorem 2.2, where x_j stands for $e_j, j \in J$ and P stands for Φ. Since Φ is isotone and condition (c) of Theorem 2.2 is clearly accomplished, application of the latter theorem yields the existence of a linear positive operator T that verifies

$$T(e_j) = y_j, j \in J, \qquad T(h) \leq \Phi(h) \ \forall h \in E, \tag{3.1}$$

To obtain the last assertion of point (a), observe that the isotonicity of Φ and (3.1) lead to

$$\pm T(h) = T(\pm h) \leq \Phi(\pm h) \leq \Phi(|h|) \Rightarrow$$

$$|T(h)| \leq \Phi(|h|) \Rightarrow \|T(h)\| \leq \|\Phi(|h|)\| \leq$$

$$\|\Phi\|\|h\|, h \in E \Rightarrow \|T\| \leq \|\Phi\|$$

and the proof is done. $\qquad\qquad\qquad\square$

The next results refer to the Markov moment problem on the space $C(K)$, where K is a compact Hausdorff topological space. All the linear solutions T appearing in the sequel are Markov operators.

Theorem 3.3. *Let K be a compact Hausdorff topological space, μ a positive regular Borel measure defined on the class of Borel subsets of $K, C(K)$ the Banach lattice of all real valued continuous functions on $K, \{\varphi_j\}_{j \in J}$ a family of linearly independent elements in $C(K), \{y_j\}_{j \in J}$ a given family of elements in $L_\mu^\infty(K)$. The following statements are equivalent*

(a) *there exists a linear (positive) bounded operator $T: C(K) \to L_\mu^\infty(K)$ such that*

$$T(\varphi_j) = y_j, j \in J, T(\varphi) \leq \left(\sup_{t \in K} \varphi(t)\right)\P \ \forall \varphi \in C(K) \tag{3.2}$$

In particular, the following equalities hold

$$T(\P) = \P, \|T\| = 1;$$

(b) *for any finite subset* $J_0 \subset J$ *and any* $\{\alpha_j; j \in J_0\} \subset \mathbb{R}$, *the following relation holds true*

$$\sum_{j \in J_0} \alpha_j y_j \leq \sup_{t \in K}\left(\sum_{j \in J_0} \alpha_j \varphi_j(t)\right)\P \qquad (3.3)$$

Proof. The implication $(a) \Rightarrow (b)$ is obvious, thanks to the properties of T. To prove $(b) \Rightarrow (a)$, one applies Theorem 2.2, implication $(c) \Rightarrow (a)$, for

$$E = C(K), F = L_\mu^\infty(K), \quad P(\psi) = \left(\sup_{t \in K} \psi(t)\right)\P, \psi \in E \qquad (3.4)$$

Observe that P defined by (3.4) is a scalar valued sublinear nondecreasing functional multiplied by the class of the constant function $\P$ in $L_\mu^\infty(K)$, hence is an isotone sublinear operator. The inequality (3.3) is equivalent to the fact that condition written at point (c) of Theorem 2.2 is accomplished. Since $L_\mu^\infty(K)$ is an order complete vector lattice, according to Theorem 2.2, there exists a positive linear operator $T: E \to F$ with the properties mentioned at point (a) of the latter theorem. In particular, it results $T(\varphi_j) = T_0(\varphi_j) := y_j, j \in J$. Moreover the following implications hold

$$\varphi \in E \Rightarrow T(\varphi) \leq \left(\sup_{t \in K}\varphi(t)\right)\P, -T(\varphi) \leq \left(\sup_{t \in K} -\varphi(t)\right)\P =$$

$$= -\left(\inf_{t \in K}\varphi(t)\right)\P \Rightarrow \left(\inf_{t \in K}\varphi(t)\right)\P \leq T(\varphi) \leq \left(\sup_{t \in K}\varphi(t)\right)\P$$

Thus (3.2) are proved. In particular, for $\varphi = \P$, it results $T(\P) = \P$. Since any $\varphi \in E$ with $\|\varphi\|_E \leq 1$ is situated in the order interval $[-\P, \P]$, the positivity of T leads to $T(\varphi) \in [T(-\P), T(\P)] = [-\P, \P] \Rightarrow \|T(\varphi)\|_F \leq 1 \Rightarrow \|T\| \leq 1$. But we have already seen that $\|T(\P)\|_F = \|\P\|_F = 1$. Hence $\|T\| = 1$ and the proof is done. $\square$

In the next theorem, K will be a compact subset of $\mathbb{R}^n$ ($n \geq 1$ is a natural number), $E = C(K)$, F an order complete Banach lattice with a strong order unit $\P_F$ such that the order interval $[-\P_F, \P_F]$ is equal to the closed unit ball of F. $j = (j_1, ..., j_n) \in \mathbb{N}^n$. $t = (t_1, ..., t_n) \in K, |j| = \sum_{k=1}^n j_k, \varphi_j(t) = t^j = t_1^{j_1} \cdots t_n^{j_n}$.

Theorem 3.4. *Let* $\{y_j; |j| \leq m\} \subset F, m \in \mathbb{N}, m \geq 1$. *The following statements are equivalent*

(a) *there exists a positive linear operator* $T: C(K) \to F$ *such that*

$$T(\varphi_j) = y_j, |j| \leq m, \left(\inf_{t \in K}\varphi(t)\right)\P_F \leq T(\varphi) \leq \left(\sup_{t \in K}\varphi(t)\right)\P_F \ \forall \varphi \in C(K), \|T\| = 1; \quad (3.5)$$

(b) *for any* $\{\beta_j; |j| \leq m\} \subset \mathbb{R}$, *the following relation holds*

$$\sum_{\substack{j\in\mathbb{N}^n \\ |j|\le m}} \beta_j y_j \le \left(\sup_{t\in K} \left(\sum_{|j|\le m} \beta_j t^j \right) \right) \mathbb{1}_F \tag{3.6}$$

Proof. One repeats the proof of Theorem 3.3, where we replace $L_{\infty,\mu}(K)$ by F, $\varphi_j(t) = t^j, t \in K, j \in J := \mathbb{N}^n$, $|j| \le m, P(\psi) = \left(\sup_{t\in K} \psi(t) \right) \mathbb{1}_A, \psi \in E = C(K)$. Some of the notations have been defined before the statement. Clearly, from (3.5) with positive and unital T, the relation (3.6) follows. For the converse, repeating the arguments from the proof of Theorem 3.3, the existence of a positive linear operator verifying (3.5) follows (via Theorem 2.2., (c)$\Rightarrow$(a)). Here the subspace $H = Sp\{\varphi_j; j \in \mathbb{N}^n, |j| \le m\}$ is the vector subspace of all polynomial functio□on K, of degree $\le m$. From (3.5), in particular, $T(\mathbb{1}_E) = \mathbb{1}_F$ follows as well and the proof is done. $\square$

When applying Mazur-Orlicz theorem (Theorem 2.4), one can work with the subspace of all polynomial functions on $K \subset \mathbb{R}^n$, without any restriction on the their degree (one proves a full Mazur-Orlicz theorem). Such a result is not a direct consequence of the density of polynomials in $C(K)$ (that could be the case of the full moment problem for $C(K), K \subset \mathbb{R}^n$). In theorem 3.4, a solution for a truncated moment problem is proposed. A linear operator T from $C(K)$ to F is called a Markov operator if T is positive and $T(\mathbb{1}) = \mathbb{1}_F$ (the definition is valid for any Hausdorff compact topological space K). It is easy to observe that a linear operator $T \in L(C(K), F)$ is a Markov operator if and only if $T(\varphi) \le \left(\sup_{t\in K} \varphi(t) \right) \mathbb{1}_F \ \forall \varphi \in C(K)$. In particular, solutions T from Theorems 3.3, 3.4 and 3.5 (the latter being proved below) are Markov operators. Let F be an order complete Banach space, having a strong order unit $\mathbb{1}_F$, $(y_j)_{j\in\mathbb{N}^n}$ a sequence in F. We prove the following theorem.

Theorem 3.5. *With the notations from Theorem 4.4, let $K = K_1 \times \cdots \times K_n \subset \mathbb{R}^n_+$ be such that $K_l \subset \mathbb{R}_+$ is compact and denote $r_l = sup K_l, l = 1, ..., n, r^j = r_1^{j_1} \cdots r_n^{j_n}, j = (j_1, ..., j_n) \in \mathbb{N}^n$. The following statements are equivalent:*
   (a) there exists a (positive) linear operator $T: C(K) \to F$ such that

$$T(\varphi_j) \ge y_j, j \in \mathbb{N}^n, \left(\inf_{t\in K} \varphi(t) \right) \mathbb{1}_F \le T(\varphi) \le \left(\sup_{t\in K} \varphi(t) \right) \mathbb{1}_F \ \forall \varphi \in C(K), \|T\| = 1;$$

   (b) $y_j \le r^j \mathbb{1}_F \ \forall j \in \mathbb{N}^n$.

Proof. The implication (a)$\Rightarrow$(b) is obvious, thanks to the properties of T. Namely, the following relations hold true

$$y_j \le T(\varphi_j) \le \left(\sup_{t\in K} \varphi_j(t) \right) \mathbb{1}_A = \left(\sup_{t\in K} (t_1^{j_1} \cdots t_n^{j_n}) \right) \mathbb{1}_A = r^j \mathbb{1}_A, \qquad j \in \mathbb{N}^n$$

To prove (b)$\Rightarrow$(a), we use the implication (c)$\Rightarrow$(a) of Theorem 2.4. The conditions mentioned at (c) of the latter theorem is accomplished, since for any finite subset $J_0 \subset \mathbb{N}^n$ the following inequalities hold

$$y_j \leq r^j \P_F, \lambda_j \geq 0, \forall j \in J_0 \Rightarrow$$

$$\sum_{j \in J_0} \lambda_j y_j \leq \left(\sum_{j \in J_0} \lambda_j r^j \right) \P_F = \left((\sum_{j \in J_0} \lambda_j t^j)|_{t=(r_1,\ldots,r_n)} \right) \P_F =$$

$$\sup_{t \in K} \left(\sum_{j \in J_0} \lambda_j t^j \right) \P_F = P\left(\sum_{j \in J_0} \lambda_j \varphi_j \right), P(\psi) := \left(\sup_{t \in K} \psi(t) \right) \P_F, \psi \in C(K)$$

According to Theorem 2.4, (c)$\Rightarrow$(a), there exists a positive linear operator $T: C(K) \to F$ with the properties mentioned at point (a) of the present theorem. The proof is complete. $\square$

Observe that for Mazur-Orlicz theorem, it is not necessary that F be a lattice (F is an order complete Banach space, which is sufficient for applying Theorem 2.4).

Theorem 3.6. *Let X be a Banach lattice, Y an order complete Banach lattice, $\{\varphi_j\}_{j \in J} \subset X_+, \{y_j\}_{j \in J} \subset Y$, G a linear positive bounded operator from X into Y, α a positive number. The following statements are equivalent*

(a) *there exists a linear positive bounded operator $F \in B_+(X,Y)$, such that*

$$F(\varphi_j) \geq y_j, \forall j \in J, F(x) \leq \alpha G(|x|), \forall x \in X, \|F\| \leq \alpha\|G\|;$$

(b) $\quad y_j \leq \alpha G(\varphi_j), \forall j \in J.$

Proof. (a)$\Rightarrow$(b) is obvious, because of $y_j \leq F(\varphi_j) \leq \alpha G(|\varphi_j|) = \alpha G(\varphi_j), \forall j \in J$. For the converse, we apply Theorem 2.4, (b)$\Rightarrow$(a). Let $J_0 \subset J$ be a finite subset, $\{\lambda_j\}_{j \in J_0} \subset \mathbb{R}_+, x \in X$, such that $\sum_{j \in J_0} \lambda_j \varphi_j \leq x$. Then using (b) and the fact that the scalars λ_j are nonnegative, as well as the positivity of G, we derive

$$\sum_{j \in J_0} \lambda_j y_j \leq \alpha \sum_{j \in J_0} \lambda_j G(\varphi_j) = \alpha G\left(\sum_{j \in J_0} \lambda_j \varphi_j \right) \leq \alpha G(x) \leq \alpha G(|x|) := P(x).$$

Application of Theorem 2.4 leads to the existence of a linear positive operator F from X into Y such that

$$F(\varphi_j) \geq y_j, \forall j \in J, F(x) \leq \alpha G(|x|), \forall x \in X.$$

From the last relation, also using the fact that the norms on X and Y are solid ($|u| \leq |v| \Rightarrow \|u\| \leq \|v\|$), we deduce

$$|F(x)| \leq \alpha G(|x|) \Rightarrow \|F(x)\| \leq \alpha\|G\|\||x|\| = \alpha\|G\|\|x\|, \forall x \in X.$$

It follows that $\|F\| \leq \alpha\|G\|$. This concludes the proof. $\qquad\square$

Corollary 3.1. *Let M be a measure space, μ a positive measure on $M, \mu(M) < \infty, X = L_\mu^p(M), 1 \leq p < \infty, g \geq 0$ an element of $L_\mu^q(M)$, where $q \in (1, \infty]$ is the conjugate of p $(1/p + 1/q = 1)$, α a positive number. Let $\{\varphi_j\}_{j \in J}, \{y_j\}_{j \in J}$ be as in Theorem 4.6, where $Y = \mathbb{R}$. The following statements are equivalent*

(a) *there exists $h \in L_\mu^q(M)$, $0 \leq h \leq \alpha g$ a.e., $\int_M h\varphi_j d\mu \geq y_j, \forall j \in J$;*

(b)$y_j \leq \alpha \int_M g\varphi_j d\mu, \forall j \in J.$

Proof. One applies Theorem 3.6 for $G(\psi) = \int_M g\psi \, d\mu$, $\psi \in X$, $Y = \mathbb{R}$, as well as the representation of linear positive continuous functionals on L^p spaces by means of nonnegative elements from L^q spaces. In order to prove (b)$\Rightarrow$(a), from the preceding results it follows that there exists $h \in \left(L_\mu^q(M)\right)_+$ such that $\int_M h\varphi_j \, d\mu \geq y_j, \forall j \in J$ and

$$\int_M h\psi d\mu \leq \alpha \int_M g\psi d\mu$$

for all nonnegative functions $\psi \in L_\mu^p(M)$. Now we choose $\psi = \chi_B$, where B is an arbitrary measurable subset of M. Then the last relation can be rewritten as

$$\int_B (h - \alpha g) \, d\mu \leq 0$$

for all such subsets B. A straightforward application of a measure theory argument leads to $h - \alpha g \leq 0$ a.e. in M. Since (a)$\Rightarrow$(b) is obvious, this concludes the proof. $\qquad\square$

Corollary 3.2. *Let consider the measure space $M = \mathbb{R}_+^n, n \in \{1, 2 \dots\}$, endowed with the measure $d\mu = exp\left(-\sum_{j=1}^n p_j t_j\right) dt_1 \cdots dt_n, p_j > 0, \forall j \in \{1, \dots, n\}$, α a positive number. The following statements are equivalent*

(a) *there exists $h \in L_\mu^\infty(\mathbb{R}_+^n), \int_{\mathbb{R}_+^n} ht^j d\mu \geq y_j, \forall j \in \mathbb{N}^n, 0 \leq h \leq \alpha$ a.e.;*

(b) $y_j \leq \alpha \dfrac{j_1! \cdots j_n!}{p_1^{j_1+1} \cdots p_n^{j_n+1}}, \forall j = (j_1, \dots, j_n) \in \mathbb{N}^n.$

Proof. One applies Corollary 3.1 to $p = 1, q = \infty, g = 1$ a.e. The notation t^j is the multi $-$ index notation $t^j = t_1^{j_1} \cdots t_n^{j_n}$. The conclusion follows via Fubini theorem and Gamma function properties. $\qquad\square$

Theorem 3.7. Let $X = L_\mu^p(M), 1 < p, \mu \geq 0, \mu(M) < \infty, \{\varphi_j\}_{j \in J} \subset X, \{y_j\}_{j \in J} \subset \mathbb{R}, \alpha > 0, \alpha \in \mathbb{R}, q$ the conjugate of p. *Consider the following statements*

(a) *there exists* $h \in \left(L_\mu^q(M)\right)_+$ *such that*

$$\int_M h\varphi_j d\mu \geq y_j, \forall j \in J, \int_M h\psi d\mu \leq \alpha \|\psi\|_p (\mu(M))^{1/q}, \qquad \forall \psi \in X;$$

(b) *we have* $y_j \leq \alpha \int_M \varphi_j \, d\mu, \forall j \in J$.

Then (b)$\Rightarrow$(a).

Proof. *Let* $J_0 \subset J$ *be a finite subset,* $\{\lambda_j\}_{j \in J_0} \subset \mathbb{R}_+$. Hölder inequality and using also (b), lead to the following implications

$$\sum_{j \in J_0} \lambda_j \varphi_j \leq \psi \Rightarrow \int_M \left(\sum_{j \in J_0} \lambda_j \varphi_j\right) d\mu \leq \int_M \psi d\mu \leq \|\psi\|_p (\mu(M))^{1/q} \Rightarrow$$

$$\sum_{j \in J_0} \lambda_j y_j \leq \alpha \int_M \left(\sum_{j \in J_0} \lambda_j \varphi_j\right) d\mu \leq \alpha \|\psi\|_p (\mu(M))^{1/q} = P(\psi).$$

Application of Theorem 2.4 and measure theory arguments yield the existence of $h \in L_\mu^q(M)$ such that

$$F(\varphi_j) = \int_M h\varphi_j d\mu \geq y_j, \forall j \in J, F(\psi) = \int_M h\psi d\mu \leq \alpha \|\psi\|_p (\mu(M))^{1/q}, \psi \in X.$$

Moreover, since $F(\psi) \geq 0, \forall \psi \in X_+$, we have

$$\int_M h\psi d\mu \geq 0, \forall \psi \in X_+.$$

Taking $\psi = \chi_B$, where $B \subset M$ is a measurable set such that $\mu(B) > 0$, one obtains

$$\int_B h d\mu \geq 0$$

for all such subsets B. Application of a well known measure theory argument leads to $h \geq 0 \, \mu - a.e.$ From the previous relations we also derive that $\|h\|_q \leq \alpha (\mu(M))^{1/q}$. This concludes the proof. $\square$

The following theorem represents an application of the general result stated in Theorem 2.4 to some other concrete spaces X, Y. Let H be an arbitrary Hilbert space, $n \in \mathbb{N}, n \geq 1, A_1, ..., A_n$

positive commuting self - adjoint operators acting on H, $(B_j)_{j \in \mathbb{N}^n}$ a sequence in Y, where $Y = Y(A_1, ..., A_n)$ is defined by

$$Y_1 := \{U \in \mathcal{A}(H); UA_j = A_jU, j = 1, ..., n\}, Y := \{V \in Y_1; UV = VU, \forall U \in Y_1\},$$
$$Y_+ = \{V \in Y; <Vh, h> \geq 0 \, \forall h \in H\}$$

Here $\mathcal{A}(H)$ is the real vector space of all self – adjoint operators. One can prove that Y is an order complete Banach lattice with respect to the usual structures induced by those defined on the real space of self – adjoint operators (cf. [13, p. 303 - 305]), and a commutative real Banach algebra. Notice that the properties of $Y = Y(A_1, ..., A_n)$, where $A_1, ..., A_n$ are as mentioned above can be proved in a similar way to those of a $Y(A)$, where A is a self – adjoint operator. Actually, one repeats the proofs from [13, p. 303 – 305], but for several commuting self – adjoint operators. Denote $A = (A_1, A_2, ..., A_n)$ and let E_A be the spectral measure attached to the n- touple $A = (A_1, A_2, ..., A_n)$ of commuting self-adjoint positive operators, and Σ_A the joint spectrum attached to A. Let denote by $\varphi_j, j \in \mathbb{N}^n$ the basic polynomials $\varphi_j(t_1, ..., t_n) = t_1^{j_1} \cdots t_n^{j_n}, j = (j_1, ..., j_n) \in \mathbb{N}^n, t = (t_1, ..., t_n) \in \Sigma_A, X := C(\Sigma_A)$.

Theorem 3.8. *The following statements are equivalent*

(a) *there exists a linear bounded positive operator $F \in B_+(X, Y)$ such that*

$$F(\varphi_j) \geq B_j, j \in \mathbb{N}^n, F(\varphi) \leq \int_{\Sigma_A} |\varphi| \, dE_{(A_1, ..., A_n)}, \forall \varphi \in X, \|F\| \leq 1;$$

(b) $B_j \leq A^j := A_1^{j_1} \cdots A_n^{j_n}, \forall j = (j_1, ..., j_n) \in J = \mathbb{N}^n.$

Proof. The implication (a)$\Rightarrow$(b) is obvious:

$$B_j \leq F(\varphi_j) \leq \int_{\Sigma_A} |\varphi_j| dE_{(A_1, ..., A_n)} = \int_{\Sigma_A} \varphi_j dE_{(A_1, ..., A_n)} = A_1^{j_1} \cdots A_n^{j_n},$$

$j \in \mathbb{N}^n$ (we have used the positivity of the operators A_k which leads to $|\varphi_j| = \varphi_j$ on Σ_A). For the converse, one applies Theorem 2.4 (b)$\Rightarrow$(a), where $\mathbb{N}^n$ stands for J, φ_j stands for x_j and B_j stands for $y_j, \forall j \in \mathbb{N}^n$. Let J_0 and $\{\lambda_j\}_{j \in J_0}$ be as mentioned at point (b) of Theorem 2.4. The following implications hold

$$\sum_{j \in J_0} \lambda_j \varphi_j \leq \varphi \in X \Rightarrow \sum_{j \in J_0} \lambda_j \int_{\Sigma_A} \varphi_j dE_{(A_1, ..., A_n)} = \sum_{j \in J_0} \lambda_j A_1^{j_1} \cdots A_n^{j_n} \leq \int_{\Sigma_A} \varphi dE_{(A_1, ..., A_n)}$$

$$\leq \int_{\Sigma_A} |\varphi| \, dE_{(A_1, ..., A_n)} := P(\varphi)$$

The positivity of the spectral measure $dE_{(A_1, ..., A_n)}$ has been used. On the other hand, the hypothesis (b), the fact that the scalars λ_j are nonnegative and the preceding evaluation yield

$$\lambda_j B_j \leq \lambda_j A^j \; \forall j \Rightarrow \sum_{j \in J_0} \lambda_j B_j \leq \sum_{j \in J_0} \lambda_j A^j = \sum_{j \in J_0} \lambda_j A_1^{j_1} \cdots A_n^{j_n} \leq P(\varphi),$$

where $T(\varphi)$ was defined above. Thus the implication at (b) Theorem 2.4 is accomplished. Application of the latter theorem leads to the existence of a "feasible solution" F having the property mentioned at point (a) of the present theorem. The last property is a consequence of the preceding one, using the fact that the norm on Y is solid. This concludes the proof. $\quad\square$

Remark 3.1. If in Theorem 3.8 one additionally assumes that $\|A_k\| < 1, k = 1,2,\dots,n$, then for any self - adjoint operators satisfying (b) one has

$$\sum_{j\in\mathbb{N}^n} B_j \leq \prod_{k=1}^{n}(I - A_k)^{-1}$$

The next theorem is an application of Theorem 2.1 [46]. It refers to a space of analytic functions as the domain space, and it is a multidimensional Markov moment problem over the real field. Let $n \neq 0$ be a natural number and X be the space of absolutely convergent power series in the unit closed polydisc $\overline{D}_1 = \{z = (z_1,\dots,z_n): |z_p| \leq 1, p \in \langle 1,\dots,n\rangle\}$, with real coefficients. The norm on X is defined by

$$\|\varphi\|_\infty = sup\{|\varphi(z)|: z \in \overline{D}_1\}.$$

Denote

$$h_k(z) = z_1^{k_1} \cdots z_n^{k_n}, k = (k_1,\dots,k_n) \in \mathbb{N}^n, z \in \overline{D}_1,$$

$|k| := k_1 + \cdots + k_n..$ On the other side, let H be a complex Hilbert space, $\mathcal{A}$ the real vector space of all self adjoint operators acting on $H, A \in \mathcal{A}$. Define the space $Y = Y(A)$ given by

$$Y_1 = \{V \in \mathcal{A}; VA = AV\}, Y = Y(A) = \{U \in Y_1; UV = VU, \forall V \in Y_1\} \qquad (3.7)$$

$$Y_+ = \{U \in Y; <Uh, h> \geq 0 \ \forall h \in H\}$$

Let $(B_k)_{k\in\mathbb{N}^n}$ be a multi - indexed sequence of operators in Y, and $\tilde{B} \in Y_+\backslash\{0\}$.

Theorem 3.9. *Assume that $A_1,\dots,A_n$ are elements of Y such that there exists a real number $M > 0$, with the property*

$$|B_k| \leq M\frac{A_1^{2k_1}}{k_1!}\cdots\frac{A_n^{2k_n}}{k_n!}, \forall k \in \mathbb{N}^n, \sum_{p=1}^{n} A_p^2 \leq I,$$

where I is the identity operator. Let $\{\varphi_k\}_{k\in\mathbb{N}^n} \subset X$ be such that $1 = \|\varphi_k\| = \varphi_k(0), \forall k \in \mathbb{N}^n$. Then there exists a linear bounded operator $F \in B(X,Y)$ such that

$$F(h_k) = B_k, |k| \geq 1, F(\varphi_k) \geq \tilde{B}, \forall k \in \mathbb{N}^n,$$

$$F(h) \leq (2 + \|\tilde{B}\|M^{-1}e^{-1})\|h\|_\infty u_0, \forall h \in X, u_0 := MeI.$$

In particular, the following evaluation holds: $\|F\| \leq 2Me + \|\tilde{B}\|$.

Proof. One applies Theorem 2.1 [46]. The subspace generated by $\{h_k \colon |k| \geq 1\}$ stands for S of Theorem 2.1 [46] and the convex hull of the set of the functions $\varphi_k, k \in \mathbb{N}^n$, stands for the set A. The following remark is essential:

$$\|s - \varphi\|_\infty \geq |s(0) - \varphi(0)| = |0 - 1| = 1, \forall s \in S, \forall \varphi \in A.$$

This proves that $\bigl(S + B(0,1)\bigr) \cap A = \emptyset$, so that $B(0,1)$ stands for V and $\|\cdot\|_\infty$ stands for p_V from Theorem 2.1. [46]. The operator $\tilde{B}$ will stand for $\tilde{y}$. Now let

$$\varphi = \sum_{j \in J_0} \beta_j h_j \in S \cap B(0,1),$$

where J_0 is a finite subset of $\mathbb{N}^n$. The following relations hold

$$\left| \sum_{j \in J_0} \beta_j B_j \right| \leq \sum_{j \in J_0} |\beta_j| |B_j| \leq \|\varphi\|_\infty \sum_{j \in J_0} \frac{1}{r_1^{j_1} \cdots r_n^{j_n}} |B_j|,$$

for any $0 < r_p < 1, p \in \{1, \dots, n\}$, thanks to Cauchy inequalities. Passing to the limit with $r_p \uparrow 1. p \in \{1, \dots, n\}$ and using the fact that $\varphi \in B(0,1)$, as well as the hypothesis in the statement, the preceding relation further yields

$$\left| \sum_{j \in J_0} \beta_j B_j \right| \leq \sum_{j \in J_0} |B_j| \leq M \sum_{j \in J_0} \frac{A_1^{2j_1}}{j_1!} \cdots \frac{A_n^{2j_n}}{j_n!} \leq M \left(\sum_{k_1 \in \mathbb{N}} \frac{A_1^{2k_1}}{k_1!} \right) \cdots \left(\sum_{k_n \in \mathbb{N}} \frac{A_n^{2k_n}}{k_n!} \right) =$$

$$= M exp \left(\sum_{p=1}^n A_p^2 \right) \leq M exp(I) = MeI = u_0.$$

The conclusion is that denoting by $f \colon S \to Y$ the linear operator which satisfies the moment conditions $f(h_k) = B_k, k \in \mathbb{N}^n, |k| > 1$, we have

$$-MeI \leq f(s) \leq MeI = u_0, \forall s \in S \cap B(0,1).$$

On the other hand, the following relations hold

$$\tilde{B} \leq \|\tilde{B}\| I = \|\tilde{B}\| M^{-1} e^{-1} u_0 = \alpha_1 u_0,$$

where $\alpha_1 := \|\tilde{B}\| M^{-1} e^{-1}$. The conditions on the norms of the functions $\varphi_k, k \in \mathbb{N}^n$ lead to

$$\|\varphi\| \leq 1, \forall \varphi \in A.$$

So, the constant 1 stands for α from Theorem 2.1 [46]. Now all the conditions from the statement of Theorem 2.1 [46] are accomplished. Application of the latter theorem, leads to the existence of a linear mapping $F \colon X \to Y$, such that

$$F(h_k) = f(h_k) = B_k, k \in \mathbb{N}^n, |k| > 1, F(\varphi_k) \geq \tilde{B}, \forall k \in \mathbb{N}^n,$$

$$F(h) \le \left(2 + \|\tilde{B}\|M^{-1}e^{-1}\right)\|h\|_\infty MeI, \forall h \in X.$$

From the last inequality, we derive

$$|F(h)| \le \left(2Me + \|\tilde{B}\|\right)\|h\|_\infty I, \qquad \forall h \in X.$$

Since the norm on Y is solid, we infer that

$$\|F(h)\| \le \left(2Me + \|\tilde{B}\|\right)\|h\|_\infty, \forall h \in X \Rightarrow \|F\| \le 2Me + \|\tilde{B}\|.$$

This concludes the proof. $\qquad\qquad\qquad\qquad\qquad\qquad\qquad\qquad\qquad\quad$ $\square$

4. Polynomial approximation on unbounded subsets and the Markov moment problem

Polynomial approximation recalled below allow proving the existence as well as the uniqueness of the solution of some classical moment problems on unbounded subsets.

Lemma 4.1. *Let $\psi : [0,\infty) \to R_+$ be a continuous function, such that $\lim\limits_{t \to \infty} \psi(t) \in R_+$ exists. Then there is a decreasing sequence $(h_l)_l$ in the linear hull of the functions*

$$\varphi_k(t) = exp(-kt), k \in \mathbb{N}, t \in [0,\infty)$$

such that $h_l(t) > \psi(t)$, $t \ge 0$, $l \in \mathbb{N}$, $\lim h_l = \psi$ uniformly on $[0,\infty)$. There exists a sequence of polynomial functions $(\tilde{p}_l)_{l \in \mathbb{N}}, \tilde{p}_l \ge h_l > \psi, \lim \tilde{p}_l = \psi$, uniformly on compact subsets of $[0,\infty)$.

Lemma 4.2. *Let v be a M-determinate positive regular measure on $[0,\infty)$, with finite moments of all natural orders. If ψ, $(\tilde{p}_l)_l$ are as in Lemma 4.1, then there exists a subsequence $(\tilde{p}_{lm})_m$, such that $\tilde{p}_{lm} \to \psi$ in $L^1_v([0.\infty))$ and uniformly on compact subsets. In particular, it follows that the positive cone P_+ of positive polynomials is dense in the positive cone $(L^1_v([0.\infty)))_+$ of $L^1_v([0.\infty))$.*

Lemma 4.3. *Let $A \subset R^n$ be an unbounded closed subset, and v an M-determinate positive regular Borel measure on A, with finite moments of all natural orders. Then for any $x \in (C_0(A))_+$, there exists a sequence $(p_m)_m$, $p_m \in P_+$, $p_m \ge x$, $p_m \to x$ in $L^1_v(A)$. In particular, we have*

$$\lim \int_A p_m(t)dv = \int_A x(t)dv,$$

P_+ is dense in $(L^1_v(A))_+$, and P is dense in $L^1_v(A)$.

Proof. Let consider the sublattice $X \subset L^1_v(A)$ of all function ψ such that $|\psi|$ is dominated by some polynomial p on A. To prove the assertions of the statement, it is sufficient to show that for any $x \in (C_0(A))_+$, we have

$$Q_1(x) := \inf\left\{\int_A p(t)dv; \ p \ge x, \ p \in P\right\} = \int_A x(t)dv$$

Obviously, one has

$$Q_1(x) \geq \int_A x(t)\, dv \qquad (4.1)$$

To prove the converse, we define the linear form

$$F_0 : X_0 = \mathcal{P} \oplus Sp\{x\} \to \mathbb{R},\ F_0(p + \alpha x) := \int_A p(t)\, dv + \alpha Q_1(x), p \in \mathcal{P}, \alpha \in \mathbb{R}$$

Next we show that F_0 is positive on X_0. In fact, for $a < 0$, one has (from the definition of Q_1, which is a sublinear functional on X)

$$p + ax \geq 0 \Rightarrow p \geq -ax \Rightarrow (-a)Q_1(x) = Q_1(-ax) \leq \int_A p(t)dv \Rightarrow F_0(p + ax) \geq 0.$$

If $a \geq 0$, we infer that

$$0 = Q_1(0) = Q_1(ax + (-ax)) \leq aQ_1(x) + Q_1(-ax) \Rightarrow$$

$$\int_A p(t)dv \geq Q_1(-ax) \geq -aQ_1(x) \Rightarrow F_0(p + ax) \geq 0.$$

Whence, in both possible cases, we have $x_0 \in (X_0)_+ \Rightarrow F_0(x_0) \geq 0$. Since X_0 contains the space of polynomials functions, which is a majorizing subspace of X, there exists a linear positive extension $F : X \to R$ of F_0 (cf. Theorem 2.5), that is continuous on $C_0(A)$, with respect to the sup-norm. Therefore, F has a representation by means of a positive Borel regular measure μ on A, such that

$$F(x) = \int_A x(t)d\mu, \quad x \in C_0(A).$$

Let $p \in \mathcal{P}_+$ be a nonnegative polynomial function. There is a nondecreasing sequence $(x_m)_m$ of continuous nonnegative function with compact support, such that $x_m \uparrow p$, pointwise on A. Positivity of F and Lebesgue dominated convergence theorem for μ yield

$$\int_A p\, dv = F(p) \geq \sup F(x_m) = \sup \int_A x_m(t)d\mu = \int_A p\, d\mu, \quad p \in P_+.$$

Thanks to Haviland theorem, there exists a positive Borel regular measure λ on A, such that

$$\lambda(p) = v(p) - \mu(p) \Leftrightarrow v(p) = \lambda(p) + \mu(p), \quad p \in P.$$

Since v is assumed to be M-determinate, it follows that

$$v(B) = \mu(B) + \lambda(B),$$

for any Borel subset B of A. From this last assertion, approximating each $x \in (L_\nu^1(A))_+$ by a nondecreasing sequence of nonnegative simple functions, and also using Lebesgue convergence theorem, one obtains firstly for positive functions, then for arbitrary ν-integrable functions x:

$$\int_A x\,d\nu = \int_A x\,d\mu + \int_A x\,d\lambda, \quad x \in L_\nu^1(A).$$

In particular, we must have

$$\int_A x\,d\nu \geq F(x) = F_0(x) = Q_1(x) \tag{4.2}$$

Now (4.1) and (4.2) conclude the proof. $\qquad\square$

Using Lemma 4.3, one can prove the following result.

Lemma 4.4. *Let $\nu = \nu_1 \times \cdots \times \nu_n$ be a product of n $M-$ determinate positive regular Borel measures on $\mathbb{R}_+ = [0,\infty)$, with finite moments of all natural orders. Then we can approximate any nonnegative continuous compactly supported function in $X = L_\nu^1(\mathbb{R}_+^n)$ by means of sums of tensor products $p_1 \otimes \cdots \otimes p_n$, p_j positive polynomial on the real nonnegative semiaxis, in variable $t_j \in [0,\infty), j = 1, ..., n$.*

Notice that a similar result holds for products of M-determinate positive regular measures on $\mathbb{R}$, with finite moments of all natural orders.

Recall that a determinate ($M-$determinate) measure is uniquely determinate by its moments, or, equivalently, by its values on polynomials. The following statement holds for any closed unbounded subset $A \subset \mathbb{R}^n$, hence does not depend on the form of positive polynomials on A. We denote by φ_j, $\varphi_j(t) := t_1^{j_1} \cdots t_n^{j_n}$, $j = (j_1, ..., j_n) \in \mathbb{N}^n, t = (t_1, ..., t_n) \in A$.

Theorem 4.1. *Let A be a closed unbounded subset of $\mathbb{R}^n$, Y an order complete Banach lattice, $(y_j)_{j \in \mathbb{N}^n}$ a given sequence in Y, ν a positive regular $M-$determinate Borel measure on A, with finite moments of all orders. Let $F_2 \in B(L_\nu^1(A), Y)$ be a linear positive bounded operator from $L_\nu^1(A)$ to Y. The following statements are equivalent*
(a) there exists a unique linear operator $F \in B(L_{1,\nu}(A), Y)$ such that $F(\varphi_j) = y_j, j \in \mathbb{N}^n$, F isbetween 0 and F_2 on the positive cone of $L_\nu^1(A)$, and $\|F\| \leq \|F_2\|$;
(b) for any finite subset $J_0 \subset \mathbb{N}^n$, and any $\{a_j\}_{j \in J_0} \subset \mathbb{R}$, we have

$$\sum_{j \in J_0} a_j \varphi_j \geq 0 \text{ on } A \implies 0 \leq \sum_{j \in J_0} a_j y_j \leq \sum_{j \in J_0} a_j F_2(\varphi_j)$$

We go on by recalling a result on the form of non-negative polynomials in a strip [26], which leads to a simple solution for the related Markov moment problem.

Theorem 4.2. *Suppose that $p(t_1, t_2) \in \mathbb{R}[t_1, t_2]$ is non $-$ negative on the strip $A = [0,1] \times \mathbb{R}$. Then $p(t_1, t_2)$ is expressible as*

$$p(t_1, t_2) = \sigma(t_1, t_2) + \tau(t_1, t_2)t_1(1 - t_1),$$

where $\sigma(t_1, t_2), \tau(t_1, t_2)$ are sums of squares in $\mathbb{R}[t_1, t_2]$.

Let $A = [0,1] \times \mathbb{R}$, v a positive $M-$ determinate regular Borel measure on A, with finite moments of all orders, $X = L_v^1(A)$, $\varphi_j(t_1, t_2) := t_1^{j_1} t_2^{j_2}, j = (j_1, j_2) \in \mathbb{N}^2, (t_1, t_2) \in A$. Let Y be on order complete Banach lattice, $(y_j)_{j \in \mathbb{N}^2}$ a sequence of given elements in Y.

Theorem 4.3. *Let $F_2 \in B_+(X, Y)$ be a linear bounded positive operator from X to Y. The following statements are equivalent*

(a) *there exists a unique bounded linear operator $F : X \to Y$, Such that*
$$F(\varphi_j) = y_j, \forall j \in \mathbb{N}^2,$$
F is between zero and F_2 on the positive cone of X, $\|F\| \le \|F_2\|$;

(b) *for any finite subset $J_0 \subset \mathbb{N}^2$, and any $\{\lambda_j; j \in J_0\} \subset \mathbb{R}$, we have*

$$0 \le \sum_{i,j \in J_0} \lambda_i \lambda_j \, y_{i+j} \le \sum_{i,j \in J_0} \lambda_i \lambda_j \, F_2(\varphi_{i+j});$$

$$0 \le \sum_{i,j \in J_0} \lambda_i \lambda_j \left(y_{i_1+j_1+1, i_2+j_2} - y_{i_1+j_1+2, i_2+j_2} \right) \le$$

$$\sum_{i,j \in J_0} \lambda_i \lambda_j \left(F_2(\varphi_{i_1+j_1+1, i_2+j_2} - \varphi_{i_1+j_1+2, i_2+j_2}) \right), i = (i_1, i_2), j = (j_1, j_2) \in J_0$$

Theorem 4.4. *Let X be as in Lemma 4.4, $(y_j)_{j \in \mathbb{N}^n}$ be a sequence in Y, where Y is an order complete Banach lattice, $F_2 \in B_+(X, Y)$. The following statements are equivalent*

(a) *there exists a unique (bounded) linear operator $F \in B(X, Y)$ such that, $F(\varphi_j) = y_j, j \in \mathbb{N}^n$„ F is between zero and F_2 on the positive cone of X, $\| F \| \le \| F_2 \|$;*

(b) *for any finite subset $J_0 \subset \mathbb{N}^n$ and any $\{\lambda_j; j \in J_0\} \subset \mathbb{R}$, we have*

$$\sum_{j \in J_0} \lambda_j \varphi_j(t) \ge 0 \ \forall t \in \mathbb{R}_+^n \Rightarrow \sum_{j \in J_0} \lambda_j y_j \in Y_+;$$

for any finite subsets $J_k \subset \mathbb{N}, k = 1, \ldots, n$, and any $\{\lambda_{j_k}\}_{j_k \in J_k} \subset \mathbb{R}, k = 1, \ldots, n$, the following relations hold

$$\sum_{i_1, j_1 \in J_1} \left(\cdots \left(\sum_{i_n, j_n \in J_n} \lambda_{i_1} \lambda_{j_1} \cdots \sum_{i_n, j_n \in J_n} \lambda_{i_n} \lambda_{j_n} \, y_{i_1+j_1+l_1, \ldots, i_n+j_n+l_n} \right) \cdots \right) \le$$

$$\sum_{i_1, j_1 \in J_1} \left(\cdots \left(\sum_{i_n, j_n \in J_n} \lambda_{i_1} \lambda_{j_1} \cdots \sum_{i_n, j_n \in J_n} \lambda_{i_n} \lambda_{j_n} F_2 \, (\varphi_{i_1+j_1+l_1, \ldots, i_n+j_n+l_n} \right) \cdots \right), (l_1, \ldots, l_n) \in \{0,1\}^n$$

Let $v = v_1 \times \cdots \times v_n$ be a product of n $M-$ determinate positive regular Borel on R, with finite moments of all natural orders. Let

$$\varphi_j(t) := t_1^{j_1} \cdots t_n^{j_n}, \; j = (j_1, \ldots, j_n) \in \mathbb{N}^n, t = (t_1, \ldots, t_n) \in \mathbb{R}^n$$

Obviously, a statement similar to that of Lemma 5.4 holds true when we replace $\mathbb{R}_+^n$ by $\mathbb{R}^n$. In the latter case, the polynomials $p_j, j = 1, \ldots, n$ are nonnegative on the whole real axis, so that they are sums of squares. Applying such an approximation result and Theorem 2.3, one obtains the following theorem.

Theorem 4.5. *Let v be as above, $X = L_v^1(\mathbb{R}^n)$, Y an order complete Banach lattice, and $(y_j)_{j \in \mathbb{N}^n}$ a multi-indexed sequence in Y. Let $F_2 : X \to Y$ be a positive linear bounded operator. The following statements are equivalent*

(a) *there exists a unique bounded linear operator $F : X \to Y$, such that $F(\varphi_j) = y_j$, $\forall j \in \mathbb{N}^n$, F is between zero and F_2 on the positive cone of X, $\| F \| \leq \| F_2 \|$;*

(b) *for any finite subset $J_0 \subset \mathbb{N}^n$, and any $\{\lambda_j; \ j \in J_0\} \subset \mathbb{R}$, we have*

$$\sum_{j \in J_0} \lambda_j \varphi_j(t) \geq 0 \; \forall t \in \mathbb{R}^n \Rightarrow \sum_{j \in J_0} \lambda_j y_j \in Y_+,$$

for any finite subsets $J_k \subset \mathbb{N}$, $k = 1, \ldots, n$, and any

$$\{\lambda_{jk}\}_{jk \in J_k} \subset \mathbb{R}, \; k = 1, \ldots, n,$$

the following relations hold

$$\sum_{i_1, j_1 \in J_1} \left(\cdots \left(\sum_{i_n, j_n \in J_n} \lambda_{i_1} \lambda_{j_1} \cdots \sum_{i_n, j_n \in J_n} \lambda_{i_n} \lambda_{j_n} \, y_{i_1 + j_1, \ldots, i_n + j_n} \right) \cdots \right) \leq$$

$$\sum_{i_1, j_1 \in J_1} \left(\cdots \left(\sum_{i_n, j_n \in J_n} \lambda_{i_1} \lambda_{j_1} \cdots \sum_{i_n, j_n \in J_n} \lambda_{i_n} \lambda_{j_n} F_2 \left(\varphi_{i_1 + j_1, \ldots, i_n + j_n} \right) \cdots \right) \right)$$

In what follows we solve an operator valued one dimensional classical Markov moment problem.

Let H be an arbitrary complex or real Hilbert space and $\mathcal{A}$ the real order vector space of all symmetric operators acting on H. The positive cone of $\mathcal{A}$ consists in all operators $U \in \mathcal{A}$, having the property: $< U(h), h > \, \geq 0 \; \forall h \in H$. Let $A \in \mathcal{A}$. Define $Y = Y(A)$ by (3.7). As is well-known, $Y(A)$ is an order complete Banach lattice. Let $X = C_{\mathbb{R}}(\sigma(A))$, where $\sigma(A) \subset [0, \infty)$ is the spectrum of the fixed positive self-adjoint operator A acting on a complex (or real) Hilbert space H. Consider the space $Y(A)$ defined above. The following corollary of the above results holds, thanks to the form of non-negative polynomials on $[0, \infty)$ [1].

Corollary 4.1. *Let $A, X, Y = Y(A)$ be as above, $(U_n)_{n \geq 0}$ be a sequence of operators in Y. The following statements are equivalent*

(a) *there exists a unique linear bounded operator $F: X \to Y$ such that the moment interpolation conditions $F(\varphi_n) = U_n, n \in \mathbb{N}$ are verified and $0 \le F(\psi) \le \psi(A), \forall \psi \in X_+, \|F\| \le 1$;*

(b) *for any finite subset $J_0 \subset \mathbb{N}$ and any $\{\lambda_j; j \in J_0\} \subset \mathbb{R}$, the following implication holds true*

$$\sum_{j \in J_0} \lambda_j t^j \ge 0, \forall t \in \sigma(A) \Rightarrow 0 \le \sum_{j \in J_0} \lambda_j U_j \le \sum_{j \in J_0} \lambda_j A^j;$$

(c) *for any finite subset $J_0 \subset \mathbb{N}$ and any $\{\lambda_j; j \in J_0\} \subset \mathbb{R}$, the following relations hold*

$$0 \le \sum_{i,j \in J_0} \lambda_i \lambda_j U_{i+j+k} \le \sum_{i,j \in J_0} \lambda_i \lambda_j A^{i+j+k}, k \in \{0,1\}$$

The last result is not necessarily related to the moment problem. However, it is characterizing the positivity of a bounded linear operator only in terms of quadratic forms. From this viewpoint, one solves the difficulty arising from the fact that nonnegative polynomials on $\mathbb{R}^n, n \ge 2$, might not be expressible as sums of squares.

Corollary 4.2. *Let X be as in Theorem 4.5, and Y be a Banach lattice. Assume that T is a linear bounded operator from X to Y. The following statements are equivalent*

(a) *T is nonnegative on the positive cone of X;*

(b) *for any finite subsets $J_k \subset \mathbb{N}, k = 1,...,n$ and any*

$$\left\{\lambda_{j_k}\right\}_{j_k \in J_k} \subset R, k = 1,...,n,$$

the following relation holds

$$0 \le \sum_{i_1, j_1 \in J_1} \left(\cdots \left(\sum_{i_n, j_n \in J_n} \lambda_{i_1} \lambda_{j_1} \cdots \lambda_{i_n} \lambda_{j_n} T\left(x_{i_1+j_1, \ldots, i_n+j_n}\right) \right) \cdots \right)$$

Proof. Notice that (b) says that T is positive on the convex cone generated by special positive polynomials similar to those mentioned in Lemma 4.4, each factor of any term in the sum being nonnegative on the whole real axis. Consequently, (a)$\Rightarrow$(b) is obvious. In order to prove the converse, observe that any nonnegative element of X can be approximated by nonnegative continuous compactly supported functions. Such functions can be approximated by sums of tensor products of positive polynomials in each separate variable. The conclusion is that any nonnegative function from X can be approximated in $X = L_\nu^1(\mathbb{R}^n)$ by sums of tensor products of squares of polynomials in each separate variable. But on such special polynomials, T admits nonnegative values, following the condition (b). Now the desired conclusion is a consequence of the continuity of T. This concludes the proof. $\square$

CHAPTER 2

EARLIER AND RECENT RESULTS ON CONVEXITY AND OPTIMIZATION

1. Introduction

The order of presentation of the results of this chapter is the chronological one, following the date of publication. Recall that solving optimization problems for convex (and concave) mappings is simpler than solving similar problems for arbitrary mappings, thanks to using Hahn-Banach type results, which are strongly related to convexity. Therefore, our results refer only to convex optimization problems. For very recent results on subdifferentials of convex operators see [31] and the references therein.

2. On convex functions defined on bounded convex subsets of $\mathbf{R}^n$

In this section, we improve the basic result from [35] (see also the references there).

Theorem 2.1. *Let X be an arbitrary real vector space, $B \subset X$ a finite dimensional convex bounded subset, Y an order complete vector lattice and $\varphi : B \to Y$ a convex operator. Then there exists $y_0 = y_0(\varphi) \in Y$ such that $\varphi(x) \geq y_0$ for all $x \in B$.*

Proof. Since B is finite dimensional and convex, its *relative interior* $ri(B)$ is nonempty. Recall that by $ri(B)$ one denotes the interior of B with respect to the topology induced on B by that on the (finite dimensional) linear variety generated by B. As is well known, φ is subdifferentiable at any point from $ri(B)$. Let $b_0 \in ri(B)$ and h a translation of a subgradient of φ at b_0, that is an affine operator $h : X \to Y$ such that $h(b_0) = \varphi(b_0)$ and $h(x) \leq \varphi(x)$ for all $x \in B$. On the other hand, let $x_1, \ldots, x_{p+1}$ (at most) $p+1$ affine independent points in the linear variety generated by B, such that

$$B \subseteq co\{x_1, \ldots, x_{p+1}\}$$

(here p is the linear dimension of the linear variety generated by B). Such a system of points does exist thanks to the fact that B is finite dimensional and bounded. Now the following relations hold

$$\varphi(x) \geq h(x) = h\left(\sum_{j=1}^{p+1} \alpha_j x_j\right) = \sum_{j=1}^{p+1} \alpha_j h(x_j) \geq \left(\sum_{j=1}^{p+1} \alpha_j\right) inf\{h(x_j); 1 \leq j \leq p+1\} =$$

$$inf\{h(x_j); 1 \leq j \leq p+1\} := y_0,$$

where $x = \sum_{j=1}^{p+1} \alpha_j x_j, \alpha_j \geq 0, j = 1, \ldots, p+1, \sum_{j=1}^{p+1} \alpha_j = 1$. This concludes the proof. $\square$

25

Remark 2.1. The preceding proof furnishes a *constructive method* of *finding a lower bound* y_0 for $\varphi(B)$.

In the next theorem, we show that the only convex subsets $B \subset X$ such that any convex real function on B is bounded from below, are the finite dimensional convex bounded subsets.

Theorem 2.2. *Let X be an arbitrary real infinite dimensional vector space and $B \subset X$ a convex subset, such that any convex real function defined on B is bounded from below. Then B is contained in a finite dimensional subspace of X and is bounded there.*

Proof. Let x^* be an arbitrary linear functional in the algebraic dual X^* of X. Then x^* and $-x^*$ are convex, and, by hypothesis, both of them are bounded from below on B. Thus $x^*(B)$ is bounded in $\mathbf{R}$. Hence B is weakly bounded in X, endowed with the weak topology corresponding to the dual pair $<X, X^*>$. Let us endow X with the finest locally convex topology which is compatible with this dual pair. By [59, Corollary 2, section 3.2, Chapter IV], we derive that B is bounded in the latter topology. Application of [59, exercise 7, Chapter II], leads to the fact that B is contained in a finite dimensional subspace and bounded there. This concludes the proof. $\qquad\qquad\square$

3.Elements of minimum norm and related results

In the sequel, we present the characterization of an element of minimum norm in terms of linear continuous forms of norm one, related to the distance function, in arbitrary normed linear spaces. Related nice geometric aspects are briefly discussed.

Lemma 3.1. *Let X be a real normed vector space, $H = \{f = \alpha\}$ a closed hyperplane in X and $x_0 \in X$. Then the distance $d(x_0, H)$ is given by formula:*

$$d(x_0, H) = \frac{|f(x_0) - \alpha|}{\|f\|}. \qquad (3.1)$$

Theorem 3.1. *Let X be a real normed vector space, $A \subset X$ a closed convex subset not containing the origin and $a_0 \in A$. The following statements are equivalent:*

(a) $\|a_0\| = \inf\{\|a\|;\ a \in A\}$;

(b) *there exists $f \in X^*$, such that*

$$\|f\| = 1, \quad \|a_0\| = f(a_0) \leq f(a), \quad \forall a \in A;$$

(c) *there exists a closed homogeneous hyperplane H such that*

$$d(H, A) = \|a_0\|.$$

Proof. Let B be the open ball of radius $\|a_0\|$, centered at the origin. From (a) we infer that the intersection of B with A is empty. From the Hahn-Banach theorem, we infer that there exists a hyperplane $H = \{f = \alpha\}$, $f \in X^* \setminus \{0\}$, which separates B from A. Scalling by a suitable constant, we can assume that $\alpha = \| a_0 \|$. Hence we have:

$$B \subset \{f < \| a_0 \|\}, \quad A \subset \{f \geq \| a_0 \|\}. \tag{3.2}$$

If $\| x \| < 1$, then we can write

$$\| \| a_0 \| \cdot x \| < \| a_0 \| \Rightarrow f(\| a_0 \| \cdot x) < \| a_0 \|.$$

This leads to $\| f \| \leq 1$. Since $a_0 \in A$, from the second inclusion (2.2) we infer that

$$f(a_0) \geq \| a_0 \| \Rightarrow f\left(\frac{a_0}{\| a_0 \|}\right) \geq 1 \Rightarrow \| f \| \geq 1.$$

The conclusion is:

$$\| f \| = 1, \quad \| a_0 \| \leq f(a_0) \leq \| f \| \cdot \| a_0 \| = \| a_0 \| \Rightarrow f(a) \geq \| a_0 \| = f(a_0), \quad \forall a \in A,$$

where we have used (3.2) once more. Now, the proof of (a) $\Rightarrow$ (b) is complete.

(b) $\Rightarrow$ (c). Let f be a functional verifying (b), $H = f^{-1}(\{0\})$. From this and also using Lemma 3.1, one obtains for any $a \in A$:

$$d(a, H) = \frac{| f(a) - 0 |}{\| f \|} = f(a) \geq \| a_0 \|, \quad \forall a \in A \Rightarrow d(A, H) \geq \| a_0 \| .$$

On the other hand,

$$d(A, H) \leq d(a_0, H) = f(a_0) = \| a_0 \|.$$

Comparing the preceding relations, we conclude that $d(A, H) = \| a_0 \|$.

(c) $\Rightarrow$ (a). This implication is almost obvious:

$$\| a_0 \| = d(A, H) = d(H, A) \leq d(0, A) \Rightarrow \| a_0 \| = \inf \{\| a \|; \ a \in A\}.$$

Now the proof of the theorem is complete. $\qquad\qquad\square$

Corollary 3.1. *Let X be a real Hilbert space, $A \subset X$ a closed convex subset not containing the origin and $a_0 \in A$.*

The following statements are equivalent:

(a) $\| a_0 \| = \inf \{\| a \|; \ a \in A\}$;

(b) $\| a_0 \|^2 \leq \langle a_0, a \rangle, \quad \forall a \in A$;

(c) $d(\{a_0\}^\perp, A) = \| a_0 \|.$

Proof. (a) $\Rightarrow$ (b). From the corresponding implication of Theorem 3.1, there exists $f \in X^*$ such that

$$\|f\| = 1, \quad \|a_0\| = f(a_0) \le f(a), \quad \forall a \in A \Rightarrow$$
$$\exists u \in X, \quad f = \langle u, \cdot \rangle, \quad \|u\| = 1, \quad \langle u, a_0 \rangle = \|a_0\| = \|a_0\| \cdot \|u\|.$$

It follows that in Cauchy-Schwarz-Buniakovski inequality occurs equality, so that we must have:

$$u = \lambda a_0 \Rightarrow \|a_0\| = \langle u, a_0 \rangle = \lambda \langle a_0, a_0 \rangle \le \langle u, a \rangle = \lambda \langle a_0, a \rangle \Rightarrow$$
$$\lambda = \frac{1}{\|a_0\|}, \quad \|a_0\|^2 \le \langle a_0, a \rangle \quad \forall a \in A.$$

The implication (b) $\Rightarrow$ (c) follows from the corresponding implication of Theorem 3.1, taking

$$H = \{a_0\}^\perp = \{u\}^\perp.$$

The implication (c) $\Rightarrow$ (a) also follows from the corresponding implication of the preceding theorem, applied for the same hyperplane $H = \{a_0\}^\perp$. $\qquad \square$

We recall the following related earlier results.

Theorem 3.2. *Let X be a normed vector space, $C \subset X$ a closed convex cone,*

$$x_0 \in X \setminus C, \quad d_0 = d(x_0, C).$$

Then there exists a linear continuous functional $f \in X^*$, such that

$$\|f\| \le 1, \quad f(u) \le 0 \ \forall u \in C, \quad f(x_0) = d_0.$$

Proof. Let $p : X \to R$ be the functional defined by $p(x) = d(x, C) = \inf\{\|x - y\|; \ y \in C\}$. Obvisouly we have $p(x_0) = d_0$, We prove that p is sublinear:

$$p(\lambda x) = \inf\{\|\lambda x - y\|; \ y \in C\} = \inf\left\{\left\|\lambda x - \lambda \cdot \frac{1}{\lambda} y\right\|; \ y \in C\right\} =$$

$$\lambda \inf\{\|x - y\|; u \in C\}, \quad \forall \lambda > 0, \quad \lambda = 0 \Rightarrow p(\lambda x) = p(0) = 0 = \lambda p(x).$$

On the other hand, for all $x_1, x_2 \in X$, $y_1, y_2 \in C$ one has:

$$\|x_1 - y_1\| + \|x_2 - y_2\| \ge \|(x_1 + x_2) - (y_1 + y_2)\| \ge p(x_1 + x_2) \Rightarrow$$
$$\inf\{\|x_1 - y_1\|; \ y_1 \in C\} + \inf\{\|x_2 - y_2\|; \ y_2 \in C\} \ge p(x_1 + x_2) \Rightarrow$$
$$p(x_1 + x_2) \le p(x_1) + p(x_2), \quad \forall x_1, x_2 \in X.$$

Let $X_0 = \{\alpha x_0; \alpha \in R\}$, $f_0 : X_0 \to R$, $f_0(\alpha x_0) = \alpha d_0$, $\forall \alpha \in R$. Then f_0 is linear and for $\alpha \ge 0$ we have

$$f_0(\alpha \cdot x_0) = \alpha \cdot d_0 = \alpha p(x_0) = p(\alpha \cdot x_0),$$

while for all $\alpha > 0$, the following remark holds

$$f_0(\alpha \cdot x_0) = \alpha \cdot d_0 < 0 \le p(\alpha \cdot x_0).$$

The conclusion is that f_0 is bounded from above by p on the one-dimensional subspace X_0. From the Hahn-Banach theorem, we infer that there exists a linear extension $f : X \to R$ of f_0, such that $f \leq p$ on X. This yields

$$f(y) \leq 0 \quad \forall y \in C, \ f(x) \leq p(x) \leq d(x,0) = \| x \|, \quad \forall x \in X,$$
$$f(x_0) = d_0 = d(x_0, C) = p(x_0).$$

Thus f is continuous, of norm at most 1, and satisfies all the other assertion from the statement. The proof is complete. $\qquad\square$

Remark. If X is a reflexive Banach space, then the distance $d(x_0, C)$ is attained at least at one point y_0 of C. In this case, we have

$$f(x_0) = \| x_0 - y_0 \| \Rightarrow f(x_0 - y_0) \geq f(x_0) = \| x_0 - y_0 \| \Rightarrow$$
$$f\left(\frac{x_0 - y_0}{\| x_0 - y_0 \|} \right) \geq 1 \Rightarrow \| f \| = 1.$$

In particular, in any Hilbert space, the linear functional from Theorem 3.2 is of norm 1.

Theorem 3.3. *Let X be a normed linear space, A,B convex subsets of X such that $d_0 = d(A,B) > 0$. Then there exists two closed parallel hyperplanes H_1, H_2 which keep the two convex subsets distanced, such that $d(H_1, H_2) = d(A,B)$.*

Proof. From the hypothesis we derive

$$d_0 = d(0, A - B) \Rightarrow B(0, d_0) \bigcap (A - B) = \Phi \Rightarrow$$
$$\exists f \in X^*, \ f(x) < d_0 \ \forall x \in B(0, d_0), \ f(y) \geq d_0 \ \forall y \in A - B.$$

The preceding relations further yield:

$$\| f \| \leq 1, \quad \inf f(A) = \alpha \geq d_0 + \sup f(B) = \beta \Rightarrow \alpha - \beta \geq d_0.$$

Now the closed hyperplanes we are looking for are

$$H_1 = \{ f = \alpha \}, \quad H_2 = \{ f = \beta \}.$$

Since the two hyperplanes separate the convex sets A and B, we have

$$d(H_1, H_2) \leq d(A, B) \ .$$

In order to prove the converse inequality, observe that the distance beween two *parallel hyperplanes* equals the distance from a point situated on a hyperplane, to the other one. This last remark, relation $\| f \| \leq 1$ and Lemma 3.1 formula (3.1), lead to:

$$d(H_1, H_2) = d(h_1, H_2) = \frac{| f(h_1) - \beta |}{\| f \|} = \frac{\alpha - \beta}{\| f \|} \geq \alpha - \beta \geq d_0 \ .$$

This concludes the proof. $\qquad\square$

In Hilbert spaces, if the algebraic difference $A - B$ is closed, then the distance from the preceding theorem is attained at a pair of points $(a, b) \in A \times B$. If the two sets have smooth boundaries, the line joining these points is orthogonal to the tangent hyperplanes at these points, which are parallel. These remarks lead to the following result, which is useful in applications. It avoids using Lagrange multipliers in determining the distance between two suitable convex sets, and the attaining points.

Corollary 3.2. *Let X be a real Hilbert space, $p : X \to \mathbb{R}$ convex and smooth, $q : X \to \mathbb{R}$ concave and smooth, such that $q(x) < p(x)$ $\forall x \in X$. If there exist $a_1, b_1 \in X$ such that*

$$d((a_1, p(a_1)), (b_1, q(b_1))) = d(\text{graph } p, \text{graph } q) > 0,$$

Then there exists $\alpha \in \mathbb{R}$, s.t.

$$\nabla p(a_1) = \alpha \nabla q(b_1), \exists \lambda \in \mathbb{R} : (\nabla p(a_1), -1) = \lambda(a_1 - b_1, p(a_1) - q(b_1))$$

Examples

1) Let X be a separable Hilbert space, $\{e_n\}_{n \in N}$ a Hilbert base in X, and consider the hyperplane

$$H = \left\{ x; \sum_{n=0}^{\infty} \alpha_n \langle x, e_n \rangle = 1 \right\}, \quad (\alpha_n)_n \in (l^2), \quad \alpha_n > 0 \ \forall n \in N.$$

Then it is not difficult to see that the orthogonal projection of the origin on H is the vector

$$P_H(0) = \sum_{n=0}^{\infty} \frac{\alpha_n}{\left(\displaystyle\sum_{j=0}^{\infty} \alpha_j^2 \right)} \cdot e_n.$$

In particular, we have

$$d(0, H) = \frac{1}{\left(\sum_{n=0}^{\infty} \alpha_n^2 \right)^{1/2}} = d(0, B), B = H \cap \{ x \in X; \ < x, e_n > \ \geq 0 \ \forall n \in N \}$$

It follows that we can determine the orthogonal projection of the origin on the base B of the cone of all elements with nonnegative Fourier coefficients. Note that it is a reflexive cone, because the closed unit ball is weakly compact. One can prove that the base B is unbounded, because its "intersection with the coordinate axis" are the points $y_n = \alpha_n^{-1} e_n$, $n \in N$, $\| y_n \| \to \infty$.

2) Let $X = l^1$ or $L^1([0,1])$, $B = \left\{ x = (x_n)_n \in l^1; \ x_n \geq 0 \ \forall n \in N, \ \sum_{n=1}^{\infty} x_n = 1 \right\}$,

respectively

$$B = \left\{ \varphi \in (L^1)_+; \ \int_0^1 \varphi(t)\,dt = 1 \right\}.$$

In both cases we have

$$d(0,B) = \| y \|_1 = 1, \quad \forall y \in B,$$

and obviously B is bounded, convex and closed. Hence there are an infinitely many points (all the points of B) at which the distance to origin is attained. This happens because of the form of ball in l^1 - norm (respectively in $L^1 -$ norm).

4. Classes of concave mappings, related constrained inequalities and optimization

4.1 Introduction

Using minimum principle for concave functions, we prove a related constrained inequality, firstly for finite sums. The case of infinite sums is deduced from the previously mentioned inequality, passing to the limit. In the end, a similar result for a class of concave operators taking values in the positive cone of a special space of self-adjoint operators is discussed. Two related types of examples are given. We follow the results of [51].

4.2. An inequality related to a special class of concave functions

Let $n \geq 2$ be a natural number, $\{c_1, \dots, c_n\} \subset (0, \infty), h \colon [1, \infty) \to \mathbb{R}_+$ a concave continuous strictly increasing function, such that $h(1) = 0$.

Theorem 4.2.1. *The following statements are valid.*

(a) *For any $\{x_1, \dots, x_n\} \subset \mathbb{R}_+$, $\sum_{j=1}^n x_j = M > 0$, the following relation holds true*

$$\sum_{j=1}^n c_j\, h(1 + x_j) \geq \left(\min_{1 \leq j \leq n} c_j \right) h(1 + M) \tag{1}$$

If h is strictly concave, then in (1) equality occurs if and only if $j_0 \in \{1, \dots, n\}$ is such that $x_j = 0, j \in \{1, \dots, n\}\backslash\{j_0\}, x_{j_0} = M$ and $c_{j_0} = \min\limits_{1 \leq j \leq n} c_j$.

(b) *Using the same notations und hypothesis, under the weaker constraint $\sum_{j=1}^n x_j \geq M > 0$, for strictly concave and increasing function h, the same relation (1) holds, and equality occurs in the same case as that mentioned at point (a).*

Proof. (a) Define the $n - 1$ dimensional simplex

$$S_{n-1,M} := \left\{ (x_1, \dots, x_n) \in \mathbb{R}^n; \ x_j \geq 0, j \in \{1, \dots, n\}, \sum_{j=1}^n x_j = M \right\}$$

This is a simplex of vertices $e_1 := (M, 0, \dots, 0), \dots, e_n := (0, 0, \dots, 0, M)$ (the set $Ex(S_{n-1,M})$ equals the set $\{e_1, \dots, e_n\}$). $S_{n-1,M}$ is contained in the $n - 1$ dimensional linear variety

$$V_{n-1} = \{e_{j_1}\} + Span\{e_j - e_{j_1}; j \in \{1, ..., n\}\setminus\{j_1\}\}$$

for any $j_1 \in \{1, ..., n\}$. On the other hand, the function

$$g(x) = g(x_1, ..., x_n) := \sum_{j=1}^{n} c_j\, h(1 + x_j), x \in K := S_{n-1,M}$$

is concave and continuous, as a sum of n functions having these two properties. Application of the minimum principle for g leads to

$$g(x) \geq \min_{1 \leq j \leq n} g(e_j) = \min_{1 \leq j \leq n} \left(c_j h(1 + M)\right) = \left(\min_{1 \leq j \leq n} c_j\right) h(1 + M)$$

This proves the first assertion of the theorem. Assume now that h is strictly concave. Then so is g. If equality occurs in (1) for $x \in K\setminus\{e_j; j \in \{1, ..., n\}\}$, then, due to Carathéodory's theorem, there exist a subset $E_1 = \{e_{j_1}, ..., e_{j_k}\}, k \in \{1, ..., n\}$ and $\{\alpha_1, ..., \alpha_k\} \subset (0, \infty), \sum_{l=1}^{k} \alpha_l = 1$, such that $x = \sum_{l=1}^{k} \alpha_l\, e_{j_l}$. Now strict concavity of g and equality in (1) yield

$$g(x) > \sum_{l=1}^{k} \alpha_l\, g(e_{j_l}) \geq \left(\min_{1 \leq l \leq k} g(e_{j_l})\right)\left(\sum_{l=1}^{k} \alpha_l\right) \geq \min_{1 \leq j \leq n} g(e_j) =$$

$$\left(\min_{1 \leq j \leq n} c_j\right) h(1 + M) = g(x),$$

which is a contradiction. This concludes the proof the proof of (a). To prove (b), consider the set

$$A_M := \left\{(x_1, ..., x_n) \in \mathbb{R}^n; \; x_j \geq 0, j \in \{1, ..., n\}, \sum_{j=1}^{n} x_j \geq M\right\} = \bigcup_{\varepsilon \geq 0} S_{n-1,M+\varepsilon}$$

Then application of the results of (a) leads to

$$\min_{x \in A} g(x) = \inf_{\varepsilon \geq 0} \inf_{x \in S_{n-1,M+\varepsilon}} g(x) = \left(\min_{1 \leq j \leq n} c_j\right) \inf_{\varepsilon \geq 0} h(1 + M + \varepsilon) =$$

$$\left(\min_{1 \leq j \leq n} c_j\right) h(1 + M)$$

Observe that the minimum of g on the unbounded closed subset A_M is attained at one of the extreme points of A_M. Latter points are exactly the extreme points of $S_{n-1,M}$. The conclusion follows. $\square$

Corollary 4.2.1. *Let $n \geq 2$ be a natural number and $\{c_1, ..., c_n\} \subset (0, \infty), p_j \geq 0, j = 1, ..., n, \sum_{j=1}^{n} p_j = 1$. Then the following relation holds:*

$$\sum_{j=1}^{n} c_j\, \ln(1 + p_j) \geq \left(\min_{1 \leq j \leq n} c_j\right) \ln(2)$$

where equality occurs if and only if $p_j = 0, j \in \{1,\ldots,n\}\setminus\{j_0\}, p_{j_0} = 1$, and $c_{j_0} = \min\limits_{1 \le j \le n} c_j$ for some $j_0 \in \{1,\ldots,n\}$.

Corollary 4.2.2. *Let h be a concave continuous strictly increasing function on $[1,\infty)$, $(c_n)_{n \ge 1}$ a bounded sequence of positive numbers such that $\inf\limits_{n \ge 1} c_n > 0$. Assume that*

$$h(1+x) < x, \forall x > 0, h(1) = 0.$$

Then for any convergent series $\sum_{n=1}^{\infty} x_n = M$ of nonnegative numbers, we have

$$\sum_{n=1}^{\infty} c_n h(1+x_n) \ge \left(\inf\limits_{n \ge 1} c_n\right) h(1+M)$$

Proof. Observe the series in the left hand side of the above inequality is convergent, because of

$$\sum_{n=1}^{\infty} c_n h(1+x_n) \le \left(\sup\limits_{n \ge 1} c_n\right) \sum_{n=1}^{\infty} x_n = M\left(\sup\limits_{n \ge 1} c_n\right) < \infty$$

For each $n \ge 2$, from Theorem 2.1 we know that $\sum_{j=1}^{n} c_j\, h(1+x_j) \ge \left(\min\limits_{1 \le j \le n} c_j\right) h(1 + \sum_{j=1}^{n} x_j)$. Passing through the limit over $n \to \infty$, one obtains

$$\sum_{n=1}^{\infty} c_n h(1+x_n) = \lim_{n \to \infty}\left(\sum_{j=1}^{n} c_j\, h(1+x_j)\right) \ge \lim_{n \to \infty}\left(\min\limits_{1 \le j \le n} c_j\right) h\left(1 + \sum_{j=1}^{n} x_j\right) =$$

$$\left(\inf\limits_{n \ge 1} c_n\right) h(1+M).$$

This concludes the proof. $\qquad\qquad\square$

Theorem 4.2.2. *Let h be a strictly increasing strictly concave continuous real function on $[1,\infty)$, such that $h(1) = 0$, and the associated function $u \to u \cdot h(1+1/u)$ is strictly increasing on $(0,\infty)$.*

(a) *Let $n \in \mathbb{N}, n \ge 2, \{c_1,\ldots,c_n\} \subset (0,\infty), x_j \ge 0, j \in \{1,\ldots,n\}$ be such that $\sum_{j=1}^{n} c_j x_j = M_n$, where $M_n > 0$ is a given constant. Then we have*

$$\sum_{j=1}^{n} c_j \cdot h(1+x_j) \ge \left(\min\limits_{1 \le j \le n} c_j\right) \cdot h\left(1 + \frac{M_n}{\min\limits_{1 \le j \le n} c_j}\right)$$

and equality occurs if and only if $x = (x_1,\ldots,x_n) = \left(0,\ldots,0,\frac{M_n}{c_{j_0}},0,,0\right)$, where

$c_{j_0} = \min\limits_{1 \le j \le n} c_j$, *and the non-null component(s) of x is (are) x_{j_0}.*

(b) *If $(c_n)_{n \ge 1}$ is a bounded sequence of positive numbers with $\inf\limits_{n \ge 1} c_n > 0$, assuming that $h(1+x) < x, \forall x > 0$, then for any sequence $(x_n)_{n \ge 1}$ of nonnegative numbers such that the sum $\sum_{n=1}^{\infty} c_n x_n = M \in (0,\infty)$, we have*

$$\sum_{n=1}^{\infty} c_n h(1 + x_n) \geq \left(\inf_{n\geq 1} c_n\right) h\left(1 + \frac{M}{\inf\limits_{n\geq 1} c_n}\right)$$

Proof. To prove (a), one repeats the idea of the proof of Theorem 2.1, where we define the simplex

$$S_{n-1} := \left\{x = (x_1, \ldots, x_n) \in \mathbb{R}^n; \; x_j \geq 0, j \in \{1, \ldots, n\}, \sum_{j=1}^{n} c_j x_j = M_n\right\}$$

Then $Ex(S_{n-1}) = \{e_1, \ldots, e_n\}, e_1 := \left(\frac{M_n}{c_1}, 0, \ldots, 0\right), \ldots, e_n := \left(0, \ldots, 0, \frac{M_n}{c_n}\right),$

$$g_n(x) := \sum_{j=1}^{n} c_j \cdot h(1 + x_j), x = (x_1, \ldots, x_n) \in S_{n-1} \Rightarrow g_n(e_j) = c_j h\left(1 + \frac{M_n}{c_j}\right),$$

$$j = 1, \ldots, n$$

Due to concavity property of g, one has

$$\min_{x \in S_{n-1}} g_n(x) = \min_{1 \leq j \leq n} g_n(e_j) = \min_{1 \leq j \leq n} c_j h\left(1 + \frac{M_n}{c_j}\right) =$$

$$\left(\min_{1 \leq j \leq n} c_j\right) \cdot h\left(1 + \frac{M_n}{\min\limits_{1 \leq j \leq n} c_j}\right),$$

where the last inequality follows from the hypothesis on the function $u \to u \cdot h(1 + 1/u), u > 0$, which was assumed to be strictly increasing. The last assertion from (a) as well as that from (b) can be deduced in a similar way to the proof of Theorems 2.1 and Corollary 2.2, by means what we have already discussed of in the present proof. $\square$

Example 4.2.1. All assumptions in the statement of Theorem 2.2 are valid in the particular case when $h(t) = ln(t), t > 0$.

Example 4.2.2. The function $h(t) = t^p - 1, t \geq 0, 0 < p < 1$ satisfies all requirements of Theorem 2.2. Verifying these assertions is an elementary task. Therefore, we shall omit the proof. In particular, Theorem 2.1 can be applied for these functions taken as h.

Remark 4.2.1. The inequality of Theorem 2.2 point (a) remains valid under the weaker constraints $x_j \geq 0, j \in \{1, \ldots, n\}, \sum_{j=1}^{n} c_j x_j \geq M_n > 0$.

The next result refers to some properties of the function $\rho(u) := u \cdot ln(1 + 1/u), u > 0$. The proof will be omitted because is too elementary.

Proposition 4.2.1. (a) *The function ρ is strictly increasing, strictly concave on $(0, \infty)$, and the horizontal line of equation $y = 1$ is an asymptote for the graph of the function ρ at infinity.*

(b) *The unique fixed point $\tilde{u} \in (0, \infty)$ of the function ρ is $\tilde{u} = \frac{1}{e-1}$. The function ρ is a contraction from $[1/(e-1), \infty)$ to $[1/(e-1), 1) \subset [1/(e-1), \infty)$, of contraction constant $\rho'\left(\frac{1}{e-1}\right) =$*

$\frac{1}{e}$. *Choosing* $u_0 = 1, u_{n+1} = \rho(u_n), n \in \mathbb{N}$, *we have the following well-known relation controlling the speed of convergence* $u_n \to \tilde{u}$

$$\left| u_n - \frac{1}{e-1} \right| \le \frac{e^{-n}}{1-e^{-1}} (1 - \ln(2)), n \in \mathbb{N}$$

(c) *Denote by* $\mathcal{U}$ *the convex cone of all strictly increasing functions* $u: (0, \infty) \to (0, \infty)$, *and also denote by* $\check{\rho}: \mathcal{U} \to \mathcal{U}, \check{\rho}(u) := \rho \circ u, \left(\rho(u(x)) = u(x)\ln\left(1 + \frac{1}{u(x)}\right), x \in (0, \infty) \right)$. *Then the unique fixed point of* $\check{\rho}$ *is the constant function* $u(x) = \frac{1}{e-1}, x > 0$.

(d) *The unique self-adjoint operator* A *with the spectrum* $\sigma(A) \subset (0, \infty)$ *which verifies the equality*

$$A \cdot \ln(I + A^{-1}) = A$$

is $A = \frac{1}{e-1} I$.

(e) *The convex cone of all strictly increasing, strictly concave continuous functions from* $(0, \infty)$ *to itself is closed with respect to the operation of composition of functions. The set of all continuous strictly increasing strictly concave functions from* $(0, \infty)$ *to itself, having a common fixed point, is convex and closed with respect to composition of functions operation. Similar results hold true for the function* $\rho_p(u) = u((1 + 1/u)^p - 1), u > 0, 0 < p < 1$. *The unique fixed point of this function is* $u_p = 1/(2^{1/p} - 1) \in (0,1)$. *Observe that*

$$p \downarrow 0 \Leftrightarrow u_p \downarrow 0, \qquad p \uparrow 1 \Leftrightarrow u_p \uparrow 1$$

4.3. An operatorial variant

We start this section by recalling some known results on self-adjoint (linear) operators acting on an arbitrary complex Hilbert space H. Let $\mathcal{A}$ be the real vector space of self-adjoint operators from H to itself. Then $\mathcal{A}$ is an ordered vector space, endowed with the order relation defined by

$$U \le V \Leftrightarrow < U(h), h > \le < V(h), h >, \forall h \in H, U, V \in \mathcal{A}$$

Unfortunately, for arbitrary $U, V \in \mathcal{A}$, the supremum $\sup\{U, V\} = U \vee V$ or/and the infimum $\inf\{U, V\} = U \wedge V$ might not exist in $\mathcal{A}$. However, the following main known result holds true.

Theorem 4.3.1. *Let* $(U_n)_{n \ge 0}$ *be a monotone nondecreasing bounded above sequence of operators in* $\mathcal{A}$. *Then there exists* $U = \sup\limits_{n \ge 0} U_n$ *in* $\mathcal{A}$ *and* U *is the pointwise limit of the sequence* $(U_n)_{n \ge 0}: \lim\limits_{n \to \infty} \|U_n(h) - U(h)\| = 0, \forall h \in H.$

Obviously, a similar conclusion follows for decreasing bounded below sequences $(U_n)_{n \ge 0}$ of elements of $\mathcal{A}$.

Remark 4.3.1. A converse-type result related to Theorem 3.1 holds true in a much more general setting and its proof is obvious. Namely, let Y be an ordered vector space, which is also a topological vector space such that the positive cone Y_+ is topologically closed. If $(U_n)_{n\geq 0}$ is an increasing sequence in Y such that there exists $U = \lim_{n\to\infty} U_n \in Y$, then there exists $\sup_{n\geq 0} U_n$ and $\sup_{n\geq 0} U_n = U$.

To avoid the fact that $\mathcal{A}$ is not a vector lattice, as well as the non commutativeness of multiplication of elements from $\mathcal{A}$, for any $A \in \mathcal{A}$ one uses the construction of the following space $Y = Y(A)$.

Theorem 4.3.2. *Let $A \in \mathcal{A}, Y_1 := \{U \in \mathcal{A}; AU = UA\}, Y = Y(A) := \{V \in Y_1; VU = UV, \forall U \in Y_1\}$. Then Y is a commutative (real) Banach algebra and an order-complete Banach lattice, where*

$$|V| := sup\{V, -V\} = \sqrt{V^2}, \forall V \in Y$$

($|V|$ is equal to the positive square root of the positive self-adjoint operator V^2).

The proof of Theorem 4.3.2 can be found in [13]. Having in mind these background-type results and the above notations, we can prove the next main new theorem of this work. In the sequel, for $A \in \mathcal{A}$, the space $Y = Y(A)$ will be that defined in Theorem 4.3.2. For a continuous real function h on the spectrum $\sigma(A)$ of an operator $A \in \mathcal{A}$, we will denote also by h the mapping obtained from h by means of functional calculus attached to A.

Theorem 4.3.3. *Let $A \in \mathcal{A}$ be such that the spectrum $\sigma(A) \subset (0, \infty)$, $(T_n)_{n\geq 1}$ a sequence of elements in $Y(A)$ for which there exist $a, b \in \mathbb{R}, 0 < a < b$, with the property that the spectrums $\sigma(T_n) \subset [a, b]$, $n \in \mathbb{N}\backslash\{0\}$. Let h be a concave continuous increasing function on $[1, \infty)$ such that $0 \leq h(1 + x) \leq x, \forall x \geq 0, h(1) = 0, h(I) = \mathbf{0}$, and $\sum_{n=1}^{\infty} x_n$ a convergent numerical series of positive terms. Denote $s := \sum_{n=1}^{\infty} x_n$. Suppose there exists $L \in (0, \infty)$ such that $|h(1 + u_1 t) - h(1 + u_2 t)| \leq L|u_1 - u_2|, \forall\{u_1, u_2\} \subset \mathbb{R}_+, \forall t \in \sigma(A)$. Then the following inequality holds*

$$\sum_{n=1}^{\infty} T_n h(I + x_n A) \geq \left(\inf_{n\geq 1} T_n\right) h(I + sA) \geq a \cdot h(I + sA) \in Y_+\backslash\{0\} \tag{2}$$

($I: H \to H$ is the identity operator).

Proof. Observe the conditions on the spectrums of T_n imply:

$$T_n = \int_{\sigma(T_n)} t dE_{T_n} \geq a \int_{\sigma(T_n)} dE_{T_n} = aI, \forall n \in \mathbb{N}\backslash\{0\} \Rightarrow \inf_{n\geq 1} T_n \geq aI \in Y_+\backslash\{0\},$$

where E_U is the (positive) spectral measure attached to the self-adjoint operator $U \in Y$. Similarly, we have $T_n \leq bI$ for all natural numbers $n \geq 1$, hence $\sup_{n\geq 1} T_n \leq bI \in Y_+$. On the other hand, one has

$$h(I + x_n A) = \int_{\sigma(A)} h(1 + x_n t)\, dE_A \leq \int_{\sigma(A)} x_n\, t dE_A =$$

$$x_n \int_{\sigma(A)} t \, dE_A = x_n A, n \geq 1, n \in \mathbb{N}$$

Consequently, also using Theorem 4.3.1, we derive that the series $\sum_{n=1}^{\infty} T_n h(1 + x_n A)$ is pointwise convergent to an element of Y_+, because of

$$\sum_{n=1}^{\infty} T_n h(I + x_n A) \leq \left(\sup_{n \geq 1} T_n \right) \left(\sum_{n=1}^{\infty} x_n \right) A \leq bIsA = bsA \in Y_+$$

The next step is to prove a result similar to that from Theorem 4.2.1, in the operatorial setting. The conclusion of the present theorem will follow via a passing to the limit operation. Let $n \geq 2$. be an arbitrary natural number,

$$S_n := \sum_{j=1}^{n} x_j, \qquad S_{n-1} := \left\{ y = (y_1, \dots, y_n) \in \mathbb{R}_+^n; \sum_{j=1}^{n} y_j = S_n \right\},$$

$$g_n(y) := \sum_{j=1}^{n} T_j h(I + y_j A), y \in S_{n-1} \tag{3}$$

Obviously, for any fixed $t > 0$, the real function $u \to h(1 + ut)$ is concave on $\mathbb{R}_+$, so that $y \to \varphi(y) := h(1 + y_j t)$ is concave on S_{n-1}. Let $\{\alpha_1, \alpha_2\} \subset \mathbb{R}_+, \alpha_1 + \alpha_2 = 1, y^{(k)} \in S_{n-1}, k \in \{1,2\}$. The following relations hold

$$\tilde{\varphi}\left(\alpha_1 y^{(1)} + \alpha_2 y^{(2)}\right) := h\left(I + \left(\alpha_1 y_j^{(1)} + \alpha_2 y_j^{(2)}\right) A\right) =$$

$$\int_{\sigma(A)} h\left(1 + \left(\alpha_1 y_j^{(1)} + \alpha_2 y_j^{(2)}\right) t\right) dE_A =$$

$$\int_{\sigma(A)} h\left(\alpha_1\left(1 + y_j^{(1)} t\right) + \alpha_2\left(1 + y_j^{(2)} t\right)\right) dE_A \geq$$

$$\alpha_1 \int_{\sigma(A)} h\left(1 + y_j^{(1)} t\right) dE_A + \alpha_2 \int_{\sigma(A)} h\left(1 + y_j^{(2)} t\right) dE_A =$$

$$\alpha_1 h\left(I + y_j^{(1)} A\right) + \alpha_2 h\left(I + y_j^{(2)} A\right) = \alpha_1 \tilde{\varphi}\left(y^{(1)}\right) + \alpha_2 \tilde{\varphi}\left(y^{(2)}\right)$$

Hence $\tilde{\varphi} : S_{n-1} \to Y$ is concave. It follows that $\psi : S_{n-1} \to Y, \psi(y) := T\tilde{\varphi}(y)$ is also concave for any $T \in Y_+$ (one can multiply by a positive operator T both members of an inequality, because the product of two self-adjoint positive permutable operators is self-adjoint and positive). On the other hand, it is straightforward that a finite sum of concave operators from a convex subset of a vector space to an ordered vector space is concave. It results that the operator $g_n : S_{n-1} \to Y_+$ defined by (3) is concave. On the other hand, the set of extreme points of S_{n-1} is $Ex(S_{n-1}) = \{e_1, \dots, e_n\}$, where

$$e_1 = (s_n, 0, \ldots, 0), \ldots, e_n = (0, \ldots, 0, s_n)$$

Let $x \in S_{n-1}$. Carathéodory's theorem leads to the existence of $\{\alpha_1, \ldots, \alpha_n\} \subset \mathbb{R}_+$, $\sum_{j=1}^n \alpha_j = 1$, such that $x = \sum_{j=1}^n \alpha_j e_j$. Now we apply Jensen inequality for the concave operator g_n (which can be proved by induction, as in the case of concave real functions). Obviously, one has $h(I) = \mathbf{0} \Rightarrow g_n(e_j) = T_j h(I + s_n A), j \in \{1, \ldots, n\}$. It results

$$g_n(x) \geq \sum_{j=1}^n \alpha_j\, g_n(e_j) = \sum_{j=1}^n \alpha_j\, T_j h(I + s_n A) \geq \left(\inf_{1 \leq j \leq n} T_j \right) h(I + s_n A)$$

To conclude the proof, observe that the Lipchitz condition on $y \to h(1 + yt)$ in variable y, uniformly with respect to parameter $t \in \sigma(A)$ leads to:

$$s_n \to s \Rightarrow \|h(I + s_n A) - h(I + s A)\| =$$

$$\sup_{t \in \sigma(A)} |h(1 + s_n t) - h(1 + st)| \leq L|s_n - s| \to 0, n \to \infty$$

This proves that

$$s_n \to s \Rightarrow \lim_{n \to \infty} h(I + s_n A) = h(I + sA) \tag{4}$$

Passing to the limit, also using (4), one obtains

$$\sum_{n=1}^{\infty} T_n h(I + x_n A) = \lim_n \sum_{j=1}^n T_j h(I + x_j A) = \lim_n g_n(x) \geq$$

$$\lim_n \left(\inf_{1 \leq j \leq n} T_j \right) h(I + s_n A) \geq \left(\inf_{n \geq 1} T_n \right) \left(\lim_n h(I + s_n A) \right) =$$

$$\left(\inf_{n \geq 1} T_n \right) h(I + sA)$$

where all the limits are considered in the topology of pointwise convergence. The last relation in (2) follows too, because of $\inf_{n \geq 1} T_n \geq aI$. $\qquad \square$

Remark 4.3.1. With the notations and under hypothesis of Theorem 3.3, consider the function $u \to ln(1 + ut), u \in \mathbb{R}_+, t \in \sigma(A) \subset (0, \infty)$. Application of Lagrange theorem yields:

$$|ln(1 + u_1 t) - ln(1 + u_2 t)| \leq \sup_{\theta > 0} \max_{t \in \sigma(A)} \frac{t}{1 + \theta t} |u_1 - u_2|$$

$$\leq \|A\| |u_1 - u_2|$$

Similarly, for $0 < p < 1$, considering the function $u \to h(1 + ut) := (1 + ut)^p - 1, u \in \mathbb{R}_+, t \in \sigma(A) \subset (0, \infty)$, one obtains

$$|h(1 + u_1 t) - h(1 + u_2 t)| = |(1 + u_1 t)^p - (1 + u_2 t)^p| =$$

$$\left| \frac{p}{(1 + \theta t)^{1-p}} t(u_1 - u_2) \right| \leq p\|A\| |u_1 - u_2| = L|u_1 - u_2|,$$

where the Lipchitz constant $L := p\|A\|$ related to the variable u does not depend on the parameter $t \in \sigma(A)$. Thus both examples 4.2.1 and 4.2.2 for h are suitable for the operatorial version of the first inequality (2) proved in Theorem 4.3.3.

Corollary 4.3.1. *Under the hypothesis and using the notations from Theorem 4.3.3, the following relations hold true*

$$\sum_{n=1}^{\infty} T_n ln(I + x_n A) \geq asA(I + sA)^{-1} \geq \frac{as\omega}{1 + s\|A\|} I$$

where $\omega := \inf \sigma(A) > 0$.

Proof. Observe that for any $t \in \sigma(A)$ one has

$$ln(1 + st) > st(1 + st)^{-1}, \qquad st \leq s \cdot \sup \sigma(A) = s\|A\|$$

We have $\sigma(I + sA) = 1 + s\sigma(A) \subset [1 + s\omega, 1 + s\|A\|]$. In particular, 0 is not an element of $\sigma(I + sA)$, so that $I + sA$ is invertible. Moreover, the following relations hold

$$\sigma((I + sA)^{-1}) = (\sigma(I + sA))^{-1} \subset \left[\frac{1}{1 + s\|A\|}, \frac{1}{1 + s\omega}\right]$$

Integrating with respect to the spectral measure E_A one obtains

$$ln(I + sA) = \int_{\sigma(A)} ln(1 + st)dE_A \geq \int_{\sigma(A)} st(1 + st)^{-1} dE_A = sA(I + sA)^{-1}$$

Thus, the first inequality in the statement follows from Theorem 3.3. For the second one, observe that

$$st(1 + st)^{-1} \geq s\omega(1 + s\|A\|)^{-1}, \forall t \in \sigma(A) \Rightarrow sA(I + sA)^{-1} \geq \frac{s\omega}{1 + s\|A\|} I$$

This concludes the proof. $\qquad\qquad\qquad\qquad\qquad\qquad\qquad\qquad\qquad\qquad\square$

5. A constrained optimization problem related to Markov moment problem

This Section starts by recalling briefly one of the earlier extension type results [37] and, on the other hand, by formulating one main problem due to Douglas Todd Norris' PhD Thesis, entitled "Optimal Solutions to the L_infinity Moment Problem with Lattice Bounds" [32], directed by Professor Emeritus Robert Kent Goodrich. The latter work suggested us the results of this section. One proves a result in a general setting, motivated by a similar problem to that considered in [32] (theorem 5.2 from below). A constrained related optimization problem in infinite dimensional spaces is solved too. The next result refers to the abstract moment problem [37], and is based on constrained extension theorems for linear operators. It was recalled in Chapter 1 from above and will be applied in the sequel. The results of this section were published in [49].

Theorem 5.1. *Let $\tilde{X}$ be a preordered vector space with its positive cone $\tilde{X}_+$, Y an order complete vector lattice, $\{x_j\}_{j\in J} \subset \tilde{X}$, $\{y_j\}_{j\in J} \subset Y$ given families, $U_1, U_2 \in L(\tilde{X}, Y)$ two linear operators. The following statements are equivalent*

(a) *there exists a linear operator $U \in L(\tilde{X}, Y)$ such that*

$$U_1(x) \le U(x) \le U_2(x), \quad \forall x \in \tilde{X}_+, \quad U(x_j) = y_j, \quad \forall j \in J;$$

(b) *for any finite subset $J_0 \subset J$ and any $\{\lambda_j\}_{j\in J_0} \subset R$, we have:*

$$\left(\sum_{j\in J_0} \lambda_j x_j = \varphi_2 - \varphi_1, \; \varphi_1, \varphi_2 \in \tilde{X}_+ \right) \Rightarrow \sum_{j\in J_0} \lambda_j y_j \le U_2(\varphi_2) - U_1(\varphi_1).$$

In particular, using the latter theorem, one obtains a necessary and sufficient condition for the existence of a feasible solution (see theorem 5.2 from below). Under such condition, the existence of an optimal feasible solution follows too. On the other hand, the uniqueness and the construction of the optimal solution seems to be not obtained easily by such general methods. Therefore, we focus mainly on the existence problem. For other aspects of such problems on an optimal solution (uniqueness or non – uniqueness, construction of a unique solution, etc.), see [2]. In the latter work, one considers the following primal problem (P)

$$v = \inf \left\{ \|\varphi\|_\infty; \varphi \in L^\infty_\mu(X), \int_X \varphi \varphi_j d\mu = b_j, j = 1, \dots, n, 0 \le \alpha \le \varphi \le \beta \right\}$$

where α, β are in $L^\infty_\mu(X)$, $\{\varphi_j\}_{j=1}^n$ is a subset of $L^1_\mu(X)$ and $b = (b_1, b_2, \dots, b_n)^T \in \mathbf{R}^n$. The function φ is unknown, and in general it is not determined by a finite number of moments. The next theorem generalizes some of the above existence – type results for a feasible solution. Here (X, S) is a measure space endowed with a σ – finite positive measure μ, and S is the σ – algebra of all measurable subsets of X.

Theorem 5.2. *Let $p \in [1, \infty)$ and q be the conjugate of p. Let $\{\varphi_j\}_{j\in J}$ be an arbitrary family of functions in $L^p_\mu(X)$, where the measure μ is σ – finite, and $\{b_j\}_{j\in J}$ a family of real numbers. Assume that $\alpha, \beta \in L^q_\mu(X)$ are such that $0 \le \alpha \le \beta$. The following statements are equivalent:*

(a) *there exists $\varphi \in L^q_\mu(X)$ such that $\int_X \varphi \varphi_j d\mu = b_j, j \in J, 0 \le \alpha \le \varphi \le \beta$;*

(b) *for any finite subset $J_0 \subset J$ and any $\{\lambda_j\}_{j\in J_0} \subset \mathbb{R}$, the following implication holds*

$$\sum_{j\in J_0} \lambda_j \varphi_j = \psi_2 - \psi_1, \quad \psi_1, \psi_2 \in (L_\mu^p(X))_+ \ \Rightarrow\ \sum_{j\in J_0} \lambda_j b_j \le \int_X \psi_2 \beta \, d\mu - \int_X \psi_1 \alpha \, d\mu$$

Moreover, the set of all feasible solutions φ (satisfying the conditions (a)) is weakly compact with respect the dual pair (L^p, L^q) and the inferior

$$v := \inf\left\{\| \varphi \|_q : \varphi \in L_\mu^q(X), \ \int_X \varphi \varphi_j \, d\mu = b_j, \ j \in J, \ 0 \le \alpha \le \varphi \le \beta\right\} \ge \| \alpha \|_q$$

is attained at an optimal feasible solution φ_0 at least.

Proof. Since the implication (a) $\Rightarrow$ (b) is obvious, the next step consists in proving that (b) $\Rightarrow$ (a). Define the real valued linear positive (continuous) forms U_1, U_2 on $\tilde{X} := L_\mu^p(X)$, by

$$U_1(\varphi) := \int_X \varphi \alpha \, d\mu, \quad U_2(\varphi) := \int_X \varphi \beta \, d\mu, \quad \varphi \in \tilde{X} .$$

Then condition (b) of the present theorem coincides with condition (b) of theorem 6.1. A straightforward application of the latter theorem, leads to the existence of a linear form U on $\tilde{X}$, such that the interpolation conditions $U(\varphi_j) = b_j$, $j \in J$ are verified and

$$\int_X \psi \alpha \, d\mu \le U(\psi) \le \int_X \psi \beta \, d\mu, \quad \psi \in \tilde{X}_+ .$$

In particular, the linear form U is positive on $\tilde{X} = L_\mu^p(X)$, and this space is a Banach lattice (in particular, $\tilde{X}$ is a complete metric topological vector space and an ordered vector space, whose positive cone $\tilde{X}_+$ is closed and generating). It is known that on such spaces, any linear positive functional is continuous (cf. [58], ch. V, sect. 5). The conclusion is that U can be represented by means of a nonnegative element $\varphi \in L_\mu^q(X)$. From the previous relations, we infer that

$$\int_X \psi \alpha \, d\mu \le \int_X \psi \varphi \, d\mu \le \int_X \psi \beta \, d\mu, \quad \psi \in \tilde{X}_+ .$$

Writing these relations for $\psi = \chi_B$, where B is an arbitrary measurable set of positive measure $\mu(B)$, one deduces

$$\int_B (\varphi - \alpha) d\mu \ge 0, \quad \int_B (\beta - \varphi) d\mu \ge 0, \quad B \in S, \ \mu(B) > 0 .$$

Then a standard measure theory argument shows that $\alpha \le \varphi \le \beta$ a.e. This finishes the proof of (b) $\Rightarrow$ (a). To prove the last assertion of the theorem, observe that the set of all feasible solutions is weakly compact by Alaoglu's theorem (it is a weakly closed subset of the closed ball centered at the origin, of radius $\| \beta \|_q$). On the other hand, the norm of any normed linear space is lower weakly semi - continuous. The conclusion is that the norm $\|\cdot\|_q$ is weakly lower semi-continuous on the weakly (convex) and compact set described at point (a), so that it attains its minimum at a function φ_0 of this set. Hence, there exists at least one optimal feasible solution. This concludes the proof. $\qquad\qquad\square$

Remark 5.1. If the set $\left\{\varphi_j\right\}_{j\in J}$ is total in the space $L^p_\mu(X)$, then the set of all feasible solutions is a singleton, so that there exists a unique solution.

CHAPTER 3

ON NEWTON's METHOD FOR CONVEX FUNCTIONS AND OPERATORS AND ITS RELATIONSHIP WITH CONTRACTION PRINCIPLE

General results on a version of global Newton's method for convex increasing or decreasing functions and operators, as well as afferent examples and applications are recalled. Connection to contraction principle is discussed in detail and applied to approximation of $A^{1/p}$, where A is a positive invertible symmetric operator acting on a finite dimensional Hilbert space and $p > 1$ is a real number. Two numerical examples for 2×2 symmetric matrices with real coefficients are given. Some other nonlinear matrix or scalar equations are solved approximately. The strength of the method consists in its global character, while the weakness is that is applicable only for convex functions and operators.

1. Introduction

The aim of this review paper is to emphasize a global Newton's like method, which works only for increasing (or decreasing) convex functions and operators. The theoretical results are recalled and numerous examples are illustrating the way of applying them. The connection to contraction principle plays a central role. Namely, we approximate the positive root $A^{1/p}$ of an arbitrary symmetric operator A acting on the Hilbert space H, with the spectrum $\sigma(A) \subset (0,\infty)$, when $p \in (1,\infty)$ (see [3]). The approximation of the solution under attention is done by means of contraction principle, applied to a contraction operator (defined by means of Newton's method), of contraction constant $(p-1)/p$. If U is the positive solution under attention, then clearly U verifies the equation $P(U) := U^p - A = \mathbf{0}$. Some related numerical examples are given as well. Recall that Newton's method can be used to approximate iteratively the solution of the equation

$$P(x) = 0$$

where P is a function or operator satisfying certain conditions (see the following results and examples). In some cases, the iteration $x_{n+1} = \varphi(x_n), n \in \mathbb{N}$ of Newton's method can be done by means of a contraction mapping φ. In such cases, the evaluations of the norms of the errors given by contraction principle could be smaller than those ensured by Newton's method. For example, if p from the above example is close to 1, then the second approximation of the solution furnished by contraction principle is good enough (see the last result and the last example of this article). Being given a Hilbert space H, and a symmetric linear operator A from H to H, the construction of a commutative algebra $Y = Y(A)$, (which is also an order complete Banach lattice) of symmetric operators acting on H, is studied in detail in [13]. This space of symmetric operators (in particular of symmetric matrices) plays a central role in the present work. The rest of this article is organized as follows. Section 2 mentions briefly the methods applied in the sequel. Section 3 is devoted to the main known results on the subject and is divided in subsections. Section 4 concludes the paper.

2. Methods

The following methods are used along this chapter

1) Newton method for convex increasing (or decreasing) functions and operators.
2) Elements of theory of symmetric operators acting on a Hilbert space. Namely, let H be an arbitrary Hilbert space and A a symmetric (linear) operator acting on H. Define

$$Y_1 = \{V \in \mathcal{S}; AV = VA\}, Y = Y(A) = \{U \in Y_1; UV = VU \; \forall V \in Y_1\}$$

$$Y_+ = \{U \in Y; \; <Uh, h> \geq 0 \; \forall h \in H\}$$

Then Y is clearly a commutative real algebra of symmetric operators. It is also an order complete real Banach lattice (for details, see [15]). Here $\mathcal{S}$ is the real ordered space of all symmetric operators acting on H.

3) Contraction principle and related successive approximation method.

3. Main Text

3.1. General type results

Let X be a $\sigma-$ order complete vector lattice, endowed with a solid $\left(|x| \leq |y| \Rightarrow \|x\| \leq \|y\|\right)$ and $o-$ continuous norm $\left(x_n \to_{in \; order} x \Rightarrow \|x_n - x\| \to 0\right)$. Let Y be a normed vector space, endowed with an order relation defined by a closed convex cone. For $a, b \in X, a < b$, we denote $[a,b] = \{x \in X, a \leq x \leq b\}$. Pet $P \in C^1([a,b], Y)$. In most of our applications, we have

$X = Y$, where Y is an order complete Banach lattice of selfadjoint operators, that is also a commutative algebra (see [15], p. 303-305) and Section 2 above).

Theorem 3.1 *Additionally assume that for each* $x \in [a,b], \exists\, [P'(x)]^{-1} \in L_+(Y,X)$

and that $a \le x \le b \Rightarrow P'(a) \le P'(x) \le P'(b).$ *If* $P(a) < 0,\ P(b) > 0,$ *then there exists a unique*

solution x^* *of the equation* $P(x) = 0,$ *where*

$$x^* := \inf x_k = \lim x_k,\ x_0 := b,\ x_{k+1} = x_k - [P'(x_k)]^{-1}[P(x_k)],\ k \in N. \tag{1}$$

Moreover, we have

$$a < x^* < b,\quad \left\| x_k - x^* \right\| \le \left\| [P'(a)]^{-1} \right\| \cdot \left\| P(x_k) \right\| \to 0. \tag{2}$$

Proof. Using induction upon k, we prove that

$$P(x_k) \ge 0,\ x_{k+1} \le x_k,\quad k \in N. \tag{3}$$

The last relations (1) and the convexity of P, yield

$$P(x_0) = P(b) > 0,\ P(x_{k+1}) \ge P(x_k) + [P'(x_k)](x_{k+1} - x_k) = P(x_k) - P(x_k) = 0$$

Hence $P(x_k) \ge 0\ \forall k \in N.$ These relations lead to

$$x_{k+1} - x_k = -[P'(x_k)]^{-1}[P(x_k)] \le 0 \Rightarrow x_{k+1} \le x_k \quad \forall k \in N.$$

We derive the following useful relations

$$P(a) < 0,\ -P(x_k) \le 0 \Rightarrow 0 \ge [P'(x_k)]^{-1}(P(a) - P(x_k)) \ge$$
$$[P'(x_k)]^{-1}([P'(x_k)](a - x_k)) = a - x_k \Rightarrow x_k \ge a,\quad \forall k \in N.$$

Using the hypothesis on the space X, there exists x^* defined by the first relations (1), and from (3) we infer that the sequence $(x_k)_k$ is decreasing. Passing through the limit in the recurrence relations (1) one obtains

$$[P'(x^*)]^{-1}[P(x^*)] = 0 \Rightarrow P(x^*) = 0.$$

From the assumptions on the positivity of $P(b),\ -P(a),$ and from the definition of x^*, we infer that $a < x^* < b.$ In order to prove (2), one uses the convexity once more

$$P(x_k) = P(x_k) - P(x^*) \ge P'(x^*)(x_k - x^*) \ge P'(a)(x_k - x^*) \Rightarrow$$
$$[P'(a)]^{-1}(P(x_k)) \ge x_k - x^* \ge 0 \Rightarrow \left\| x_k - x^* \right\| \le \left\| [P'(a)]^{-1} \right\| \cdot \left\| (P(x_k)) \right\|,\ k \in N.$$

The uniqueness of the solution follows quite easily:

$$0 = P\!\left(x_1^*\right) - P\!\left(x_2^*\right) \ge P'\!\left(x_2^*\right)\!\left(x_1^* - x_2^*\right) \Rightarrow 0 \ge \left[P'\!\left(x_2^*\right)\right]^{-1}\!\left(P'\!\left(x_2^*\right)\!\left(x_1^* - x_2^*\right)\right) = x_1^* - x_2^*.$$

Similarly, we can write: $0 \ge x_2^* - x_1^*$, hence $x_1^* = x_2^*$. $\qquad\qquad\square$

The corresponding statement for convex decreasing operators holds.

Theorem 3.2 *Assume that for any $x \in [a,b]$ there exists $[P'(x)]^{-1}$ such that $-[P'(x)]^{-1}(Y_+) \subset X_+, a \le x \le b \Rightarrow [P'(x)]^{-1} \ge [P'(b)]^{-1}$. If $P(a) > 0$, $P(b) < 0$, then there exists an unique solution $x^* \in\,]a,b[$ of the equation $P(x) = 0$, x^* being given by*

$$x^* = \sup x_k = \lim x_k,\; x_0 = a,\; x_{k+1} = x_k - \left[P'(x_k)\right]^{-1}\!\left(P(x_k)\right),\quad k \in N.$$

Moreover, the sequence $(x_k)_k$ is increasing and the convergence rate is given by the inequalities

$$\left\| x_k - x^* \right\| \le \left\| [P'(b)]^{-1} \right\| \cdot \left\| P(x_k) \right\|,\, k \in N.$$

3.2. Direct consequences

During this Section we mention some applications of the general theorems of Section 2. The difficulties consist only in technical details concerning verifying conditions from general theorems. That is why we do not prove all the statements.

Theorem 3.3 *Let H be a finite dimensional Hilbert space and $X = Y$ the commutative algebra defined in [13], p. 303-305. Let*

$$B_j \in X_+,\; j \in \{0,1,...,n\},\, B_0 > 0,\, B_n > 0$$

be such that

$$B_0 < \sum_{j=1}^{n} B_j$$

and $nB_n U^{n-1} + \cdots + 2B_2 U + B_1$ is invertible for all $U \in [0,I]$. Then there exists a unique $\overline{U}, 0 < \overline{U} < I$, such that

$$B_n \overline{U}^n + \cdots + B_1 \overline{U} - B_0 = 0,$$

and this solution verifies in particular the relation

$$\left\| I - \overline{U} \right\| \le \left\| \sum_{j=1}^{n} B_j - B_0 \right\|$$

Proof. The space X is an order-complete Banach lattice and a commutative algebra of symmetric operators. Let

$$P:[0,I]\subset X_+ \to X$$

be defined by

$$P(U) = B_n U^n + \cdots + B_1 U - B_0$$

One can show that the operator $P_n(U)=U^n$ is convex on X_+. Now it follows easily that P is also convex on X_+, since all the coefficients $B_k \in X_+$ and all the operators in X are permutable. On the other hand, we have

$$P'(U)(V) = (nB_n U^{n-1} + \cdots + 2B_2 U + B_1)V \Rightarrow$$

$$[P'(U)]^{-1}(\tilde{V}) = (nB_n U^{n-1} + \cdots + 2B_2 U + B_1)^{-1}\tilde{V}, \forall \tilde{V} \in X,\ U \in [0,I]$$

We also have

$$[P'(U)]^{-1} \geq 0, 0 \leq U \leq I \Rightarrow P'(0)V = B_1 V \leq$$

$$(nB_n U^{n-1} + \cdots + 2B_2 U + B_1)V \leq (nB_n + \cdots + 2B_2 + B_1)V \Rightarrow$$

$$P'(0) \leq P'(U) \leq P'(I)\ \ \forall U \in [0,I]$$

Due to the hypothesis, we infer that

$$P(0)=-B_0 < 0, \quad P(I)= \sum_{k=1}^{n} B_k - B_0 > 0,$$

so that all requirements of Theorem 3.1 are accomplished. It follows that there exists a unique solution $\overline{U} \in]0,I[$ of the equation $P(U)=0$, that verifies the following relation (for $k = 0$ in Theorem 3.1)

$$\|I - U\| \leq \left\|[P'(0)]^{-1}\right\| \cdot \|P(I)\| = \|B_1^{-1}\| \cdot \|B_n + \cdots + B_1 - B_0\|$$

Now the proof is complete. $\qquad\qquad\qquad\qquad\qquad\qquad\qquad\qquad\qquad\qquad\ \square$

Theorem 3.4. *Let* H, X *be as in the preceding theorem,* $\alpha > 1, B \in X$ *such that the spectrum*

$S(B)\subset]\ln(\alpha),\infty[$. *There is a unique solution* $\overline{U} \in]0,I[\subset X$ *of the equation*

$$\exp(BU) - \alpha I = 0$$

and this solution verifies in particular the relation

$$\|I - \overline{U}\| \le \|B^{-1}\| \cdot \|\exp B - \alpha I\|.$$

The next result is an application of the scalar version of Theorem 3.2, when $X = Y = R$.

Proposition 3.1 *Let* $\alpha, \beta, \gamma > 0$ *be such that*

$$1 - \alpha \cdot \exp(-\alpha) - \beta < \gamma < 1.$$

Then there exists a unique solution $x^* \in \,]0,1[$ *of the equation*

$$\exp(-\alpha x) - \beta x - \gamma = 0$$

and we have

$$0 < x^* \le \frac{1 - \gamma}{\alpha \exp(-\alpha) + \beta} \to 0, \quad \gamma \uparrow 1.$$

Theorem 3.5. *Let* H *be a finite dimensional Hilbert space, and* X *the space defined in* [15], *p. 303-305. Let* $\tilde{A}, B, C \in X$ *be such the spectrums of* $\tilde{A}$ *and* B *are contained in* $\,]0, \infty[$. *Assume also that*

$$\exp(-\tilde{A}) - B < C < I.$$

Then there exists a unique solution $\overline{U} \in \,]0, I[$ *of the equation*

$$\exp(-\tilde{A}U) - BU - C = 0$$

and the following estimation holds true

$$\|\tilde{U}\| \le \left\|\left[\tilde{A}\exp(-\tilde{A}) + B\right]^{-1}\right\| \cdot \|I - C\| \to 0, C \to I$$

Proposition 3.2. *There is a unique solution* $x^* \in \,]3/2, 2[$ *of the equation*

$$2x^3 - 4x^2 + 1 = 0,$$

and this solution verifies

$$2 - \frac{27}{152} < x^* < 2.$$

Theorem 3.6. *Let* A *be a symmetric operator acting on a finite dimensional Hilbert space, with the spectrum* $S(A) \subset [3/2, 2]$. *Let* $X = Y = X(A)$ *e the space defined in Section 2. Then there exists a unique operator* $\overline{U} \in X$ *such that*

$$2\overline{U}^3 - 4\overline{U}^2 + I = 0$$

and the spectrum of this operator verifies the following relation

$$S(\overline{U}) \subset]2 - \frac{27}{152}, 2[$$

3.3. Approximating $A^{1/p}$, $p > 1$; connection to contraction principle

Let H be a finite dimensional Hilbert space, A a symmetric operator acting on H, with the spectrum $S(A) \subset]0, \infty[$, $X = X(A)$ the associated commutative algebra according to [15], p. 303-305. We denote

$$\omega_A = \inf_{\|h\|=1} \langle Ah, h \rangle, \quad \Omega_A = \sup_{\|h\|=1} \langle Ah, h \rangle$$

Theorem 3.7. *Let A be as above, $A \notin Sp\{I\}$, $p > 1$, $p \in R$. There exists a unique operator $U_p \in]\omega_A^{1/p} I, \Omega_A^{1/p} I[$ such that*

$$U_p^p - A = 0,$$

and this solution verifies the relations

$$\left\| \Omega_A^{1/p} I - U_p \right\| \leq \frac{1}{p \omega_A^{(p-1)/p}} \left\| \Omega_A I - A \right\|, \left\| U_p - \omega_A^{1/p} I \right\| \leq \frac{\Omega_A^{(p+1)/p}}{p} \left\| \omega_A^{-1} I - A^{-1} \right\|.$$

Corollary 3.1 *With the above notations and assumptions, we have*

$$\ln \Omega_A - \ln \omega_A \leq \frac{1}{\omega_A} \left\| \Omega_A I - A \right\| + \Omega_A \left\| \omega_A^{-1} I - A^{-1} \right\|.$$

Remark 3.1 If in the recurrence relation of Newton's method

$$x_{k+1} = \varphi(x_k), \; \varphi(x) := x - [P'(x)]^{-1}(P(x))$$

the mapping φ is a contraction, the rate of convergence of the sequence $(x_k)_k$ is given by contraction principle. Next, we show that this is the case of the operator $P(U) = U^p - A$, which leads to the positive solution $U_p = A^{1/p}$.

Theorem 3.8. (see also [3]). *Let p, A, X be as above. Then the Newton recurrence for the equation*

$$P(U) = U^p - A = 0$$

is

$$U_0 = \Omega_A^{1/p} I, \quad U_{k+1} = \varphi(U_k) = \frac{p-1}{p} U_k + \frac{1}{p} U_k^{-p+1} A, \quad k \in N.$$

The convergence rate for $U_k \to A^{1/p}$ is given by

$$\left\| U_k - A^{1/p} \right\| \le \left(\frac{p-1}{p} \right)^k \Omega_A^{1/p} \left\| I - \Omega_A^{-1} A \right\|, \quad k \in N.$$

Proof. Newton's sequence for the convex operator P is

$$U_0 = b = \Omega_A^{1/p} I, \, U_{k+1} = U_k - \left(pU_k^{p-1} \right)^{-1} \left(U_k^p - A \right) =$$

$$U_k - \frac{1}{p} U_k^{-p+1+p} + \frac{1}{p} U_k^{-p+1} A = \frac{p-1}{p} U_k + \frac{1}{p} U_k^{-p+1} A = \varphi(U_k), \, k \in N.$$

Let

$$M := \left\{ U \in X; U \ge A^{1/p} \right\}_{,}$$

here the root is obtained by the aid of functional calculus for A. Clearly, M is closed in X, hence it is complete. Let $\varphi : M \to X$ be defined by

$$\varphi(U) = \frac{p-1}{p} U + \frac{1}{p} U^{-p+1} A, \quad U \in M.$$

A straightforward computation shows that $\varphi\left(A^{1/p} \right) = A^{1/p}$. First we show that

$$S(U) \subset]0, \infty[\Rightarrow \varphi(U) \in M;$$

(in particular this proves that $\varphi(M) \subset M$). One can show that φ is convex on the subset of all operators in X having the spectrum contained in the positive semiaxis. In particular, φ is convex on M. Direct computations yield

$$\varphi(U) \ge \varphi\left(A^{1/p} \right) + \varphi'\left(A^{1/p} \right)\left(U - A^{1/p} \right) = A^{1/p}.$$

Thus $\varphi(U) \in M$ for all U with spectrum $S(U) \subset]0, \infty[$. Now we prove that $\varphi : M \to M$ is a contraction, with contraction constant $q = \frac{p-1}{p}$. Precisely we prove that

$$\left\| \varphi'(U) \right\| \le \frac{p-1}{p}, \quad \forall U \in M.$$

Indeed, we have

$$\|\varphi'(U)\| = \frac{p-1}{p}\|I - U^{-p}A\|; U \in M \Rightarrow U^p \geq A \Rightarrow I - U^{-p}A \geq 0 \Rightarrow$$

$$\|I - U^{-p}A\| \leq \|I\| = 1 \Rightarrow \|\varphi'(U)\| \leq \frac{p-1}{p}, \quad U \in M.$$

Now the conclusion on φ being a contraction follows by a standard differential calculus argument. Application of contraction theorem and an elementary computation shows that

$$\left\|U_k - A^{1/p}\right\| \leq \frac{q^k}{1-q}\left\|U_0 - \varphi(U_0)\right\| = \left(\frac{p-1}{p}\right)^k \Omega_A^{1/p}\left\|I - \Omega_A^{-1}A\right\|, \quad k \in N.$$

The proof is complete. $\qquad\qquad\qquad\qquad\qquad\qquad\qquad\qquad\qquad\qquad\square$

Numerical examples

1) We approximate $\begin{pmatrix} 2 & 2 \\ 2 & 5 \end{pmatrix}^{1/(\ln 6)}$, evaluating the norm of the error. Notice that $\ln(6)$ is not an integer. Therefore, simple recurring approximating sequences of the matrix from above might be difficult to find. We apply the previous Theorem 3.8. Consider the linear symmetric operator A defined by the matrix $\begin{pmatrix} 2 & 2 \\ 2 & 5 \end{pmatrix}$, applying $\mathbf{R}^2$ onto itself. The *spectrum* of A is $\sigma(A) = \{1,6\} \subset]0,\infty[$, and A is not contained in $\text{Span}\{I\}$, so that all conditions of theorem 3.8 are accomplished. Applying the latter theorem, one obtains the following two approximations of $A^{1/(\ln 6)}$:

$$U_0 = \Omega_A^{1/(\ln 6)} = 6^{1/(\ln 6)}I = eI,$$

$$U_1 = \frac{\ln 6 - 1}{\ln 6}U_0 + \frac{1}{\ln 6}U_0^{1-\ln 6}A = \frac{e}{6\cdot(\ln 6)}\begin{pmatrix} 6\ln 6 - 4 & 2 \\ 2 & 6\ln 6 - 1 \end{pmatrix}.$$

For the evaluation of the norm of the error corresponding to the second approximation U_1, one applies the last relation in the statement of Theorem 3.8, where $k = 1$, $p = \ln 6$. One deduces

$$\|U_1 - A^{1/(\ln 6)}\| \leq \frac{\ln 6 - 1}{\ln 6}6^{1/(\ln 6)}\|I - (1/6)A\| =$$

$$= \left(1 - \frac{1}{\ln 6}\right)e\|B\|, \quad B := I - \left(\frac{1}{6}\right)A = \begin{pmatrix} 2/3 & -1/3 \\ -1/3 & 1/6 \end{pmatrix}.$$

The spectrum of the matrix B is $\sigma_B = \{0, 5/6\} \subset [0,\infty[$ and B is symmetric, so that $\|B\| = \Omega_B = \frac{5}{6}$. Thus

$$\| U_1 - A^{1/(\ln 6)} \| \le \left(1 - \frac{1}{\ln 6}\right) e \frac{5}{6} \le \left(1 - \frac{1}{1.8}\right) \cdot 2.72 \cdot \frac{5}{6} = 1 + \frac{1}{135}.$$

The conclusion is

$$A^{1/(\ln 6)} = \begin{pmatrix} 2 & 2 \\ 2 & 5 \end{pmatrix}^{1/(\ln 6)} \approx \frac{e}{6 \cdot (\ln 6)} \begin{pmatrix} 6\ln 6 - 4 & 2 \\ 2 & 6\ln 6 - 1 \end{pmatrix} = U_1$$

and the norm of the error in the latter approximation is smaller than $1 + \dfrac{1}{135}$.

2) One approximates $\begin{pmatrix} 2 & 2 \\ 2 & 5 \end{pmatrix}^{1/(n^{1/n})}$, $n \in \mathbb{N}$, $n \ge 2$, evaluating the norm of the error. Using the

notations and some of the results for preceding Example 1), and applying Theorem 3.8 to the
$= = (\ )$ defined in Section 2, associated to the operator (or matrix) A, we find:

$$U_0(n) = 6^{1/(n^{1/n})} I, \quad U_1(n) = \frac{n^{1/n} - 1}{n^{1/n}} 6^{1/(n^{1/n})} I + \frac{1}{n^{1/n}} A \left(6^{1/(n^{1/n})} I\right)^{1 - n^{1/n}} =$$

$$= \frac{1}{n^{1/n}} \begin{pmatrix} (n^{1/n} - 1)6^{1/(n^{1/n})} + 2 \cdot 6^{\frac{1-n^{1/n}}{n^{1/n}}} & 2 \cdot 6^{\frac{1-n^{1/n}}{n^{1/n}}} \\ 2 \cdot 6^{\frac{1-n^{1/n}}{n^{1/n}}} & (n^{1/n} - 1)6^{1/(n^{1/n})} + 5 \cdot 6^{\frac{1-n^{1/n}}{n^{1/n}}} \end{pmatrix}.$$

Thus $\begin{pmatrix} 2 & 2 \\ 2 & 5 \end{pmatrix}^{1/(n^{1/n})} \approx U_1(n)$ and the norm of the error follows from the last evaluation in the

statement of Theorem 3.8

$$\| A^{1/(n^{1/n})} - U_1(n) \| \le \frac{n^{1/n} - 1}{n^{1/n}} 6^{1/(n^{1/n})} \cdot \left\| I - \frac{1}{6} A \right\|.$$

Applying the result from Example 1) for $\left\| I - \frac{1}{6} A \right\|$, the latter relation further yields

$$\| A^{1/(n^{1/n})} - U_1(n) \| \le \frac{5}{6^{1-1/(n^{1/n})}} \cdot \frac{n^{1/n} - 1}{n^{1/n}} \le 5(n^{1/n} - 1) \to 0, \quad n \to \infty$$

The conclusion is that the second approximation $U_1(n)$ is good enough, for large n.

CHAPTER 4

INVARIANT SUBSPACES AND INVARIANT BALLS OF BOUNDED LINEAR OPERATORS

1. Introduction

We follow the results form [45]. For the history of the invariant subspace problem and related results see the references in [45]. The first aim of the present work is to give solutions to the invariant subspace problem in a special Fréchet space of entire operator valued applications. Also, one makes the connection to the hyperinvariant subspaces of an arbitrary bounded operator, by means of an example. Finally, the invariance of the unit ball of some L^1 spaces with respect to linear bounded operators is discussed. The paper is organized as follows. Section 2 is devoted to the existence of common invariant subspaces in a Fréchet space of entire mappings. The connection with the derivation operation is briefly pointed out in Section 3. In Section 4, a related differential equation involving normal operators is discussed. An example concerning the whole first part of this work is given too. Finally, in Section 5, the invariance of the unit ball in L^1 spaces is discussed. This is the second aim of this work. To do this, polynomial approximation results are applied. Some of the characterizations in Section 5 are realized in terms of quadratic forms (with vector – coefficients).

2. Existence of invariant subspaces for special operators

If X_1, X_2 are two topological vector spaces (TVS), we denote by $LC(X_1, X_2)$ the space of all linear continuous operators applying X_1 into X_2. If $X_1 = X_2$, one denotes $LC(X_1) := LC(X_1, X_1)$. If $\mathfrak{A} \subseteq LC(X_1)$ is a nonempty subset, we denote $\mathfrak{A}^2 = \{U_1 U_2; \ U_j \in \mathfrak{A}, j = 1,2\}$.

Proposition 2.1. *Let X_1 be a TVS, such that for any $\mathfrak{A} \subseteq LC(X_1), \mathfrak{A}^2 \subseteq \mathfrak{A}$, there exists a closed proper common invariant subspace for all $U \in \mathfrak{A}$. Then any topological vector space X linearly and topologically isomorphic to X_1 has the corresponding similar property.*

Proof. Let X_1, X be as in the statement of the theorem. Consider a linear topological isomorphism $\varphi \in Isom(X, X_1)$. Let $T \in \mathfrak{X} \subseteq LC(X)$. Then the mapping $\Phi: LC(X) \to LC(X_1)$

$$\Phi(T) = \varphi \circ T \circ \varphi^{-1} = U$$

verifies

$$\Phi(T_1 T_2) = \Phi(T_1)\Phi(T_2), \forall T_j \in \mathfrak{X}, j = 1,2,$$

that is Φ is multiplicative, and so is Φ^{-1}. It follows that Φ applies any subset $\mathfrak{X} \subseteq LC(X)$, $\mathfrak{X}^2 \subseteq \mathfrak{X}$, onto a corresponding subset $\mathfrak{A} = \Phi(\mathfrak{X}) \subseteq LC(X_1)$, and the corresponding assertion on Φ^{-1} holds. Moreover, the following relations are true

$$\mathfrak{A}^2 = \left(\Phi(\mathfrak{X})\right)^2 = \Phi(\mathfrak{X}^2) \subseteq \Phi(\mathfrak{X}) = \mathfrak{A}.$$

On the other hand, by hypothesis, there exists a closed proper common invariant subspace $S \subset X_1$ of all operators $U = \Phi(T) \in \Phi(\mathfrak{X})$. Then $\tilde{S} := \varphi^{-1}(S)$ is a closed proper invariant subspace of all operators $T \in \mathfrak{X}$. $\qquad\square$

Let F be a complex Banach space and $B(F)$ the space of all linear bounded operators acting on F. As a model of a concrete Fréchet space, we consider the space

$$X_1 = E = E\big(\mathbb{C}, B(F)\big)$$

of all entire functions, having the form

$$f: \mathbb{C} \to B(F), f(z) = \sum_{n=0}^{\infty} U_n z^n, \qquad limsup \sqrt[n]{\|U_n\|} = 0 \tag{1}$$

The space E, endowed with the sequence of seminorms

$$q_n(f) = \sup_{|z| \le n} \|f(z)\|, f \in E, n \in \mathbb{N}, n \ge 1, f \in E$$

is a Fréchet space. This is a consequence of the Cauchy formula, which works for the elements of E. The topology defined by these seminorms is the topology of uniform convergence on compact subsets of $\mathbb{C}$.

Lemma 2.1. *Let* $f \in E, z \in \mathbb{C}$. *Then for any admissible contour* γ *surrounding* z, *Cauchy's formula is true:*

$$f(z) = \frac{1}{2\pi i} \int_\gamma \frac{f(\zeta)}{\zeta - z} d\zeta$$

Proof. If f is given by (1), due to the uniform convergence of the series from below on any compact subset of $\mathbb{C}$, the following equalities hold

$$f(z) = \sum_{n=0}^{\infty} U_n z^n = \sum_{n=0}^{\infty} U_n \frac{1}{2\pi i} \int_\gamma \frac{\zeta^n}{\zeta - z} d\zeta = \frac{1}{2\pi i} \int_\gamma \frac{\sum_{n=0}^{\infty} U_n \zeta^n}{\zeta - z} d\zeta = \frac{1}{2\pi i} \int_\gamma \frac{f(\zeta)}{\zeta - z} d\zeta.$$

All the series appearing above are absolutely convergent with respect to the operatorial norm on $B(F)$. This concludes the proof. $\qquad\square$

Now the fact that E is complete can be proved as in the case of complex valued functions, using Cauchy's criterion for sequences of applications with values in Banach spaces and the implication

$$f_n \to f \Rightarrow f_n' \to f' \text{ in } E.$$

This last implication is a consequence of the Cauchy formula. Also notice that the preceding result leads to the derivation term by term of a function $f \in E$, where E is given by means of (1). If $U \in B(F)$, then the operator G_U defined by

$$G_U(f)(z) = Uf(z), z \in \mathbb{C}, f \in E.$$

is an element of $LC(E)$. Namely, the following evaluation holds true

$$q_n\big(G_U(f)\big) \le \|U\| q_n(f), f \in E, n \in \mathbb{N}, n \ge 1.$$

Theorem 2.1. *Let* $\ \mathfrak{U}^2 \subseteq \mathfrak{U} \subseteq B(F)$. *Then there exists closed proper common invariant subspaces in* E, *of all operators* $G_U, U \in \mathfrak{U}$.

Proof. Define $S_0 \subset E$ as the vector subspace generated by all applications having the form

$$Uf(zL), U \in \mathfrak{U}, f \in E_0, L \in \mathfrak{L}, z \in \mathbb{C},$$

where E_0 is a nonempty subset of entire complex functions vanishing at the origin, $\mathfrak{L}$ is an arbitrary nonempty subset of $B(F)$. Now let $\tilde{U}$ be an arbitrary element of $\mathfrak{U}$, and $g \in S_0$,

$$g(z) = \sum_k c_k U_k f_k(zL_k), c_k \in \mathbb{C}, U_k \in \mathfrak{U}, f_k \in E_0, L_k \in \mathfrak{L}, z \in \mathbb{C}.$$

Then

$$G_{\tilde{U}}(g)(z) = \sum_k c_k \tilde{U}U_k f_k(zL_k), z \in \mathbb{C},$$

where the sum is finite, $\tilde{U}U_k \in \mathfrak{U}$, so that one deduces that $G_{\tilde{U}}(S_0) \subseteq S_0$. Since we have already remarked that $G_{\tilde{U}}$ is continuous on E, we infer that $G_{\tilde{U}}(\bar{S}_0) \subseteq \bar{S}_0$. It remains to prove that $\bar{S}_0 \ne E$, as we now show. Notice that the convergence in E implies the pointwise convergence, so that any element of $\bar{S}_0$ is vanishing at the origin, as being the limit of a sequence of functions having this property. On the other hand, the hyperplane H_0 of all functions in E vanishing at the origin is closed in E, by the same reason. The conclusion is that the common closed subspace $\bar{S}_0$, which is invariant for all the elements $G_{\tilde{U}}, \tilde{U} \in \mathfrak{U}$, is contained in the closed hyperplane H_0. Hence, it is a closed proper common invariant subspace. This concludes the proof. $\qquad\square$

Corollary 2.1. *Let* $U_0 \in B(F)$. *Then there exist common closed proper invariant subspaces of all operators* G_U, *where* $U \in B(F)$ *is such that* U *commutes with* U_0.

Proof. One applies Theorem 2.1 to the algebra $\mathfrak{U}$ of all operators commuting with U_0. Obviously, the equality $\mathfrak{U}^2 = \mathfrak{U}$ holds. $\qquad\square$

Corollary 2.2. *There exists common proper closed invariant subspaces of all operators* G_U, *when* $U \in B(F)$.

Proof. One applies Theorem 2.1 to $\mathfrak{U} = B(F)$. $\qquad\qquad\qquad\qquad\qquad\qquad\square$

3. Connections with classical operations

Let $U \in B(F)$ *be an invertible operator,* $S = \{f \in E; f(z) = U^k exp(zU), z \in \mathbb{C}, k \in \mathbb{Z}\}, \tilde{S} = conv(S)$ *be the convex hull of* S. *Let* $D \in LC(E)$ *be the derivation operator and* $J(g)(z) = U^{-1}g(z), z \in \mathbb{C}, g \in \tilde{S}$. *J stands for an integration operation.*

Remark 3.1. *The following assertion holds true:*

$$D(\bar{g}) = G_U(\bar{g}), J(\bar{g}) = G_{U^{-1}}(\bar{g})$$

for all applications $\bar{g}$ *in the topological closure of* $\tilde{S}$.

In fact, a straightforward calculation shows that $D = G_U$ on $\tilde{S}$. Since both operators D, G_U are continuous on $\tilde{S}$, their equality on the topological closure of $\tilde{S}$ follows. Obviously, $D = J^{-1}$ on $\tilde{S}$, and the equality can be extended to the closure of $\tilde{S}$. The last equality in the statement is obvious.

4. A related differential equation

In the next result we give a new result involving normal operators. Let H be an arbitrary Hilbert space and $T \in B(H)$ a normal operator. Consider the differential equation

$$T^n Y^{(n)} + T^{n-1} Y^{(n-1)} + \cdots + TY' - Y = 0 \qquad\qquad (2)$$

where $Y : \mathbb{C} \to B(H)$ is the unknown mapping.

Theorem 4.1. *The following statements are equivalent*

(a) $Y(z) = exp(\omega z T^*), \omega > 0$ *is a solution of the differential equation (2);*
(b) *there exists a unitary operator* $U \in B(H)$ *such that*

$$T = (\omega^{-1} t_n)^{1/2} U,$$

where t_n *is the unique positive root of the algebraic equation*

$$P(t) := t^n + t^{n-1} + \cdots + t - 1 = 0 \qquad\qquad (3)$$

Proof. In order to prove (a)$\Rightarrow$(b), we compute the left hand side member of (2), using the hypothesis. One deduces

$$(T^n \omega^n T^{\star n} + \cdots + T\omega T^{\star} - I)exp(\omega z T^{\star}) = 0.$$

Multiplying with $exp(-\omega z T^{\star})$ and using the fact that T is normal, the last relation can be rewritten as

$$(\omega T T^{\star})^n + \cdots + (\omega T T^{\star}) = I.$$

Let $A_1 = (TT^{\star})^{1/2}$ be the positive square root of the self - adjoint positive operator $TT^{\star}$. The last relation yields

$$P_1(\omega A_1^2) = (\omega A_1^2)^n + \cdots + (\omega A_1^2) = I.$$

This equality further yields

$$P_1\big(\sigma(\omega A_1^2)\big) = \sigma\big(P_1(\omega A_1^2)\big) = \sigma(I) = 1.$$

Since all the operators involved in the last equality are self - adjoint, one deduces that

$$P(t) = P_1(t) - 1 = 0, \forall t \in \sigma(\omega A_1^2),$$

where P is defined by (3). Since $\sigma(\omega A_1^2)$ consists only in positive numbers, and the equation

$$P(t) = 0$$

has a single positive root t_n we infer that

$$\sigma(\omega A_1^2) = \{t_n\},$$

where t_n is the positive root of (3).

The last equality shows that

$$\omega A_1^2 = t_n I.$$

Following the preceding notations, this may be rewritten as

$$(\omega t_n^{-1})^{1/2} T (\omega t_n^{-1})^{1/2} T^{\star} = I,$$

or, equivalently, $UU^{\star} = U^{\star}U = I$, where $U = (\omega t_n^{-1})^{1/2} T, T^{\star} = (\omega^{-1} t_n)^{1/2} U^{\star}$. Thus (a)$\Rightarrow$(b) is proved. Conversely, assume that (b) holds. We have

$$T^n Y^{(n)} + T^{n-1} Y^{(n-1)} + \cdots + T Y' - Y =$$

$$\left((\omega^{-1} t_n)^{\frac{n}{2}} U^n \omega^n (\omega^{-1} t_n)^{\frac{n}{2}} U^{\star n} + \cdots + (\omega^{-1} t_n)\omega U U^{\star} - I\right) exp\left((\omega t_n)^{\frac{1}{2}} z U^{\star}\right) =$$

$$= P(t_n) exp\left((\omega t_n)^{\frac{1}{2}} z U^{\star}\right) = 0 \qquad \square$$

Example 4.1. The vector space S of all the solutions of (2) is invariant with respect to the derivation operation and hyperinvariant with respect to G_T. This assertion holds true since the derivation - operator commutes with any operator not depending on z. Consider the space E of all entire functions from the complex plane into $B(H)$, endowed with the topology of uniform convergence on the compact subsets of $\mathbb{C}$. This is a Fréchet space, on which the derivation operation is continuous (from E onto E). Hence, the kernel of the linear differential operator involved in (2) is a closed subspace of E, so that the subspace S of all the solutions of (2) is closed. In order to formalize the above assertion, for $U \in B(H)$ let $G_U : E \to E$ be the linear continuous operator defined by $G_U(f)(z) = Uf(z), z \in \mathbb{C}$. If U commutes with T and $Y \in S$, one deduces

$$T^n\big(G_U(Y)\big)^{(n)} + \cdots + T\big(G_U(Y)\big)' - G_U(Y) =$$

$$= T^n(UY)^{(n)} + \cdots + T(UY)' - UY = U\big(T^n Y^{(n)} + \cdots + TY' - Y\big) = 0.$$

Hence $G_U(Y) \in S$ for all $Y \in S$ and all operators $U \in B(H)$ commuting with T. Notice that this example does not depend on the proof of Theorem 2. It works for an arbitrary, not necessarily normal operator $T \in B(H)$.

Remark 2. If $n = 2$, we have $t_2 = \left(-1 + \sqrt{5}\right)/2$.

5. On the invariance of the unit ball in L^1 spaces

We recall the following polynomial approximation result on closed unbounded subsets.

Theorem 5.1. *Let $A \subset R^n$ be an unbounded closed subset and v a positive M - determinate regular Borel measure on A, with finite moments of all orders. Then for any function $\psi \in (C_0(A))_+$, there is a sequence $(p_m)_m$ of polynomials on A, $p_m \geq \psi$, $p_m \to \psi$ in $L_v^1(A)$. We have*

$$\lim \int_A p_m dv = \int_A \psi \, dv,$$

the cone $\mathcal{P}_+$ of positive polynomials is dense in $\left(L_v^1(A)\right)_+$ and $\mathcal{P}$ is dense in $L_v^1(A)$.

A complete proof of this theorem was given firstly in [42]. See also Chapter 1 above. Recall that a M determinate measure is, by definition, uniquely determinate by its moments, or, equivalently, by its values on polynomials. Let Y be an order–complete Banach lattice. Let $X = L_v^1(A)$, where A, v are as in Theorem 5.1, $\mathcal{P}$ being the subspace of polynomials on A. Define the set

$$S_1^+ = \{T \in B(X, Y); T(\varphi) \geq 0 \; \forall \varphi \in X_+, \|T\| \leq 1\}.$$

Theorem 5.2. *For a linear operator $V : \mathcal{P} \to Y$, the following statements are equivalent*

(a) *V has a linear positive extension $\tilde{V} \in S_1^+$;*
(b) *there exists $T \in S_1^+$ such that*

$$0 \leq V(p) \leq T(p), \qquad \forall p \in \mathcal{P}_+.$$

Proof. The implication (a)$\Rightarrow$(b) is obvious: put $T := \tilde{V} \in S_1^+$. Then $V(p) = \tilde{V}(p) = T(p), p \in \mathcal{P}$. In order to prove the converse, consider a linear positive extension $\bar{V}$ of V, to the subspace of X formed by all functions dominated on A, in absolute value, by a polynomial. The latter space contains both the subspace of polynomials and the subspace of continuous compactly supported functions. The existence of such an extension follows from [15] or [60] (see also

Chapter 1 from above). Let ψ, p_m, $m \in \mathbb{N}$ be as in Theorem 5.1, where ψ is compactly supported. Assume by reduction to absurd that

$$T(\psi) - \bar{V}(\psi) \notin Y_+$$

Since the positive cone Y_+ is closed and convex, a separation Hahn–Banach result implies the existence of a linear positive functional $y^\star$ on Y such that

$$y^\star\big(T(\psi) - \bar{V}(\psi)\big) < 0,$$

that is

$$y^\star\big(T(\psi)\big) < y^\star\big(\bar{V}(\psi)\big).$$

On the other hand, since all the polynomials p_m are majorizing the nonnegative function ψ, using Fatou's lemma for the linear positive functional $y^\star \circ \bar{V}$, which can be represented by a positive measure, the following relations hold true

$$y^\star\big(\bar{V}(\psi)\big) \le liminf(y^\star \circ \bar{V})(p_m) = liminf(y^\star \circ V)(p_m) \le liminf(y^\star \circ T)(p_m)$$

$$= \lim_m (y^\star \circ T)(p_m) = y^\star\big(T(\psi)\big).$$

Hence we have been leaded to the contradiction

$$y^\star\big(T(\psi)\big) < y^\star\big(T(\psi)\big).$$

The conclusion is that

$$T(\psi) - \bar{V}(\psi) \in Y_+,$$

that is $\bar{V}(\psi) \le T(\psi)$ for all nonnegative continuous compactly supported functions ψ. Now let φ an arbitrary continuous compactly supported function. Then by the preceding relations, the following inequalities hold too

$$|\bar{V}(\varphi)| \le \bar{V}(|\varphi|) \le T(|\varphi|).$$

Since the norm on the Banach lattice Y is solid, we infer that

$$\|\bar{V}(\varphi)\| \le \|T\| \|\varphi\|_1 \le \|\varphi\|_1.$$

Hence $\bar{V}$ is linear, positive, continuous and of norm at most one on the dense subspace of X formed by the continuous compactly supported functions. By a standard density argument, it has a unique linear extension $\tilde{V}$ to the whole space X, of norm at most one. This bounded extension is also positive on X_+ due to the density of positive polynomials in X_+. This concludes the proof. $\qquad\qquad\square$

Denote

$$S_1^+(X) := \big\{T \in B_+(X); T(\bar{B}_{1,X}) \subseteq \bar{B}_{1,X}\big\},$$

where X is as above, and $\bar{B}_{1,X}$ is the closed unit ball in X. Let $V: P \to X$ be a linear operator.

Corollary 5.1. *The following statements are equivalent*

(a) *V has a linear positive extension $\tilde{V} \in S_1^+(X)$;*
(b) *there exists $T \in S_1^+(X)$ such that*

$$0 \leq V(p) \leq T(p), \qquad \forall p \in \mathcal{P}_+.$$

Proof. Put in Theorem 5.2 $Y = X = L_v^1(A)$. Then X is an order complete vector lattice (in which any order convergent sequence is convergent in the norm topology cf. [58]). To prove (b)$\Rightarrow$(a), one applies the corresponding implication from Theorem 5.2. Observe also that

$$\| \tilde{V} \| \leq 1 \text{ if and only if } \tilde{V}(\bar{B}_{1,X}) \subseteq \bar{B}_{1,X}.$$

This concludes the proof. $\qquad\qquad\qquad\qquad\qquad\qquad\qquad\qquad\qquad\quad$ $\square$

Our next goal is to give some characterizations in terms of quadratic forms (when this fact is allowed by the form of positive polynomials by means of sums of squares).

Corollary 5.2. *Let $X = L_v^1(\mathbb{R})$, where v is a positive regular $M -$ determinate measure on $\mathbb{R}$ (with finite moments of all orders), $x_j(t) = t^j, t \in \mathbb{R}, j \in \mathbb{N}$. Let $V: \mathcal{P} \to X$ be a linear operator. The following statements are equivalent*

(a) *V has a linear positive extension $\tilde{V} \in S_1^+(X)$;*
(b) *there exists $T \in S_1^+(X)$ such that for any finite subset $\{\lambda_j\}_{j \in J_0} \subset \mathbb{R}$, the following relations hold*

$$0 \leq \sum_{i,j \in J_0} \lambda_i \lambda_j V(x_{i+j}) \leq \sum_{i,j \in J_0} \lambda_i \lambda_j T(x_{i+j})$$

Proof. One applies Corollary 5.1 to $X = L_v^1(\mathbb{R})$, when in Theorem 5.1 one takes $n = 1, A = \mathbb{R}$, also using the form of positive polynomials on the real line, as being sums of squares of some polynomials with real coefficients [1]. $\qquad\qquad\qquad\qquad\qquad$ $\square$

Corollary 5.3. *Let $X = L_v^1([0, \infty)), v$ being a positive regular measure on $\mathbb{R}_+$, with finite moments of all orders. Let $x_j(t) = t^j, t \in \mathbb{R}_+, j \in \mathbb{N}$. Let $V: \mathcal{P} \to X$ be a linear operator. The following statements are equivalent*

(a) *V has a linear positive extension $\tilde{V} \in S_1^+(X)$;*
(b) *there exists $T \in S_1^+(X)$ such that for any finite subset $\{\lambda_j\}_{j \in J_0} \subset \mathbb{R}$, the following relations hold*

$$0 \leq \sum_{i,j \in J_0} \lambda_i \lambda_j V(x_{i+j+l}) \leq \sum_{i,j \in J_0} \lambda_i \lambda_j T(x_{i+j+l}), l \in \{0,1\}.$$

Proof. The proof is similar to that of Corollary 5.2., also using the form of positive polynomials on $\mathbb{R}_+$ [2]: $p(t) \geq 0 \ \forall t \in \mathbb{R}_+$ if and only if there exist $p_1, p_2 \in \mathbb{R}[t]$ such that $p(t) = p_1^2(t) + tp_2^2(t), \ \forall t \geq 0.$ $\qquad\qquad\square$

As it is well known, in several dimensions, there are positive polynomials which are not sums of squares cf. [4]. However, using approximation results, the connection with tensor products of positive polynomials in each separate variable holds true. Namely, one can prove the following approximation lemma, based on Theorem 5.1 from above (see also Chapter 1 from above).

Lemma 5.1. *Let* $v = v_1 \times \cdots \times v_n$ *be a product of* $n \geq 2$ $M-$ *determinate positive regular Borel measures on* $\mathbb{R}$, *with finite moments of all natural orders. Then we can approximate any nonnegative continuous compactly supported function in* $L_v^1(\mathbb{R}^n)$ *by means of sums of tensor products* $p_1 \otimes \cdots \otimes p_n$ p_j *positive polynomial on the real line, in variable* t_j, $j = 1,...,n.$

Let $X = L_v^1(\mathbb{R}^n), n \geq 2, v$ be as in Lemma 5.1.

$$x_j(t_1, ..., t_n) = t_1^{j_1} \cdots t_n^{j_n}, j = (j_1, ..., j_n) \in \mathbb{N}^n, (t_1, ..., t_n) \in \mathbb{R}^n$$

Let $V: \mathcal{P} \to X$ be a linear operator, where $\mathcal{P}$ is the subspace of polynomials in n real variables, with real coefficients.

Theorem 5.3. *The following statements are equivalent*

    (a) *V has a linear positive extension* $\tilde{V} \in S_1^+(X)$;
    (b) $0 \leq V(p) \ \forall p \in \mathcal{P}_+$ *and there exists* $T \in S_1^+(X)$ *such that for any finite subsets* $J_k \subset \mathbb{N}, k = 1,...,n$ *and any*

$$\left\{ \lambda_{j_k} \right\}_{j_k \in J_k} \subset R, k = 1,...,n,$$

we have

$$\sum_{i_1, j_1 \in J_1} \left(\cdots \left(\sum_{i_n, j_n \in J_n} \lambda_{i_1} \lambda_{j_1} \cdots \lambda_{i_n} \lambda_{j_n} V\left(x_{i_1+j_1, \ldots, i_n+j_n} \right) \right) \cdots \right) \leq$$

$$\sum_{i_1, j_1 \in J_1} \left(\cdots \left(\sum_{i_n, j_n \in J_n} \lambda_{i_1} \lambda_{j_1} \cdots \lambda_{i_n} \lambda_{j_n} T\left(x_{i_1+j_1, \ldots, i_n+j_n} \right) \right) \cdots \right)$$

Proof. Observe that (b) says that V is nonnegative on the convex cone of nonnegative polynomials, and is dominated by T on the convex cone generated by the tensor products

$$p_1 \otimes \cdots \otimes p_n$$

where each p_k is a square of a polynomial of variable $t_k \in \mathbb{R}, k = 1, \ldots, n$. In particular, such tensor products are nonnegative on $\mathbb{R}^n$. Hence (a)$\Rightarrow$(b) is obvious, putting $T = \tilde{V}$. In order to prove the converse, extend V to a linear positive operator $\bar{V}$ defined on the subspace X_1 of all functions from X whose absolute value are dominated by a polynomial (depending on that function). Then the subspace X_1 contains the polynomials and the compactly supported continuous functions too. Next we prove that $\bar{V}(\psi) \leq T(\psi)$ for all continuous nonnegative compactly supported functions ψ. By reduction to absurd, assume the contrary. Since $T(\psi) - \bar{V}(\psi) \notin X_+$, a Hahn – Banach separation theorem shows that there exists a positive (continuous) linear form $x^\star$ on X such that $x^\star\big(T(\psi) - \bar{V}(\psi)\big) < 0,$ that is

$$x^\star\big(T(\psi)\big) < x^\star\big(\bar{V}(\psi)\big) \tag{4}$$

On the other side, by a polynomial approximation result of Chapter 1, there exists a sequence

$$\sum_{j=0}^{k(m)} p_{m,1,j} \otimes \cdots \otimes p_{m,n,j} \to \psi, m \to \infty$$

in the norm topology of the space X. Now one applies Fatou's lemma for the positive functional $x^\star \circ \bar{V}$, which can be represented by means of a positive measure. Using the second relation of the point (b), one deduces

$$x^\star\big(\bar{V}(\psi)\big) \leq liminf_m (x^\star \circ \bar{V})\left(\sum_{j=0}^{k(m)} p_{m,1,j} \otimes \cdots \otimes p_{m,n,j} \right) \leq$$

$$liminf_m (x^\star \circ T)\left(\sum_{j=0}^{k(m)} p_{m,1,j} \otimes \cdots \otimes p_{m,n,j} \right) = x^\star\big(\bar{T}(\psi)\big), \psi \in \big(C_c(\mathbb{R}^n)\big)_+$$

which contradicts (4). The conclusion is that $\bar{V}(\psi) \leq T(\psi)$ for all nonnegative continuous compactly supported function ψ. Next one shows that there exists a linear extension $\tilde{V}$ of $\bar{V}$, from X into X, of norm dominated by $\|T\| \leq 1$. The positivity of $\tilde{V}$ follows via density arguments. This concludes the proof. $\qquad\square$

CHAPTER 5

FROM LINEAR TO SUBLINEAR AND TO CONVEX OPERATORS

1. Introduction

One of the aims of this work is to prove a uniformly boundedness theorem for convex continuous operators. Such results were considered in [28] and [65]. However, there are some differences between the methods and results proved in [28], [65] and our methods/results, the latter seeming to be improvements of the former ones. For example, we obtain the convexity of the neighborhood W_0 of Theorem 1 proved below, and its properties, without assuming that the domain space E is locally convex, unlike the corresponding result from [28]. Moreover, we reduce the problem to the case of symmetric convex continuous operators. The special case of uniformly bounded families of sublinear operators is discussed in detail. The second goal of this paper is to extend some inequalities involving continuous linear and sublinear operator from a small set of test points to the positive cone of the domain space. Relationship between convex and linear continuous operators is emphasized as well. The rest of the paper is organized as follows. In Section 2, a uniformly boundedness property for convex operators is proved. Some corollaries are deduced, most of them involving sublinear operators. Section 3 is devoted to extending inequalities which hold at all extreme points of a weakly compact convex subset contained in the positive cone to all points of the positive cone. One applies Krein-Milman theorem (to prove propositions 1 and 2 of Section 3).

2. Uniformly boundedness of some families of convex operators

In the sequel, E will be a (not necessarily locally convex) topological vector space which cannot be expressible as the countable union of closed subsets having empty interiors, and F will be a locally convex vector lattice (on which the lattice operations are continuous and there exists a fundamental system $\mathcal{V}$ of neighborhoods V of 0_F which are convex, closed and solid subsets, i. e.

$$|y_1| \leq |y_2|, y_2 \in V \Rightarrow y_1 \in V)$$

Both spaces E, F are vector spaces over the real field. Consider a class $\mathcal{P}$ of convex continuous operators $P: E \to F, P(0_E) = 0_F$. Recall that we can always reduce the problem of proving the equicontinuity of a family of convex operators at a point $x_0 \in E$ to the equicontinuity of a corresponding family of convex operators at 0_E, where each element P of the latter family satisfies the condition $P(0_E) = 0_F$ (cf. [28], the proof of Theorem 3.1). It is possible that some assertions of Theorem 1 from below be true under more general conditions. On the other hand, the proof of this theorem is not the same to that of Theorem 3.1 [28] (see the first comments in the Introduction and compare the two statements and their proofs).

Theorem 1. *Assume that for each $V \in \mathcal{V}$, and any $x \in E$, there exists a small enough positive number r such that*

$$rP(x) \in V \;\; \forall P \in \mathcal{P}$$

Then for any $V_0 \in \mathcal{V}$, there exists a closed convex neighborhood W_0 of 0_E such that

$$\bigcup_{P \in \mathcal{P}} P(W_0) \subset V_0$$

One writes $\lim\limits_{x \to 0_E} P(x) = 0_F$ uniformly in $P \in \mathcal{P}$.

Proof. For any $V_0 \in \mathcal{V}$ and any $P \in \mathcal{P}$, define $P_1: E \to F$, $P_1(x) := sup\{P(x), P(-x)\}$, $x \in E$. The operator P_1 is obviously convex. An additional property of P_1 is $P_1(x) = P_1(-x), x \in E$. Consequently, the codomain of P_1 is F_+, since $0_F = P_1(0_E) = P_1\left(\frac{1}{2}x + \frac{1}{2}(-x)\right) \leq \frac{1}{2}2P_1(x) = P_1(x), x \in E$. The operator P_1 is also continuous, as the least upper bound of two continuous operators, thanks to the continuity of "sup" operation from $F \times F$ to F. The subset $P_1^{-1}(V_0)$ is closed, due to the continuity of P_1. Now we prove that it is also convex. Indeed, for $x_1, x_2 \in P_1^{-1}(V_0), t \in [0,1],$. the following relations hold

$$P_1\big((1-t)x_1 + tx_2\big) \leq (1-t)P_1(x_1) + tP_1(x_2) \in V_0,$$

since V_0 is convex and P_1 is convex too. Now using assumption on V_0 of being solid, it results $P_1\big((1-t)x_1 + tx_2\big) \in V_0 (\Leftrightarrow \big((1-t)x_1 + tx_2\big) \in P_1^{-1}(V_0))$. Define

$$W_0 := \cap_{P \in \mathcal{P}} P_1^{-1}(V_0)$$

The subset W_0 is closed and convex, as an intersection of such subsets. Clearly, $\cup_{P \in \mathcal{P}} P_1(W_0) \subset V_0$. For any $x \in W_0$ and any $P \in \mathcal{P}$, it results

$$|P(x)| \leq sup\{P(x), P(-x)\} = P_1(x) \in V_0,$$

because of $-P(x) \leq P(-x), x \in E$. Indeed, $0_F = P(0_E) \leq \frac{1}{2}(P(x) + P(-x)), x \in E$. Having in mind the property of V_0, we infer that $P(x) \in V_0, \forall x \in W_0, \forall P \in \mathcal{P}$. The first conclusion is $\bigcup_{P \in \mathcal{P}} P(W_0) \subset V_0$. To finish the proof, we have to show that W_0 is a neighborhood of 0_E. For any $x \in E$ and for any $V_0 \in \mathcal{V}$, there exists a sufficiently small $r_0 > 0$ such that $\alpha P_1(x) \in V_0 \ \forall \alpha \in \mathbb{R}, |\alpha| \leq r_0, \forall P \in \mathcal{P}$. We can suppose that $r_0 \leq 1$. From the preceding considerations, it results

$$\alpha \in [0, r_0] \subset [0,1] \Rightarrow P_1(\alpha x) = P_1\big((1 - \alpha)0_E + \alpha x\big) \leq \alpha P_1(x) \in V_0 \Rightarrow P_1(\alpha x) \in V_0$$
$$\alpha \in [-r_0, 0] \Rightarrow P_1(\alpha x) = P_1\big((-\alpha)(-x)\big) \leq (-\alpha)P_1(-x) \leq r_0 P_1(x) \in V_0, P \in \mathcal{P}$$

These relations lead to: $x \in E, |\alpha| \leq r_0 \Rightarrow \alpha x \in W_0 \Rightarrow x \in \frac{1}{|\alpha|} W_0 \subset n W_0$ for sufficiently large $n \in \mathbb{N}$. Consequently, the following basic relation holds true $E = \bigcup_{n \in \mathbb{N}} n W_0$. Now recall that W_0 is closed, convex, and our assumption on E yields $int(W_0) \neq \emptyset$, so that there exists $x_0 \in int(W_0) \Rightarrow 0_E = \frac{1}{2}(x_0 + (-x_0)) \in int(W_0)$. This concludes the proof. $\quad\square$

Corollary 1. *Let E be a Banach space, F a Banach lattice, $\mathcal{P}$ a collection of continuous convex operators $P \colon E \to F, P(0_E) = 0_F$, such that for any $x \in E$, we have $\sup_{P \in \mathcal{P}} \|P(x)\|_F < \infty$. Then the following relation holds:* $\sup_{P \in \mathcal{P}, \|x\|_E \leq 1} \|P(x)\|_F < \infty.$

In the sequel, E will be an (F) space, i.e. a metrizable complete (not necessarily locally convex) topological vector space, F will be a normed vector lattice (in particular, its norm is monotone on $F_+ \colon (0_F \leq y_1 \leq y_2 \Rightarrow \|y_1\|_F \leq \|y_2\|_F)$ and the multiplication with scalars is continuous). Recall that a normed vector lattice F is a vector lattice endowed with a solid norm $(|y_1| \leq |y_2| \Rightarrow \|y_1\|_F \leq \|y_2\|_F)$, such that the lattice operations are continuous. Consider a class $\mathcal{S}$ of sublinear operators $\Phi \colon E \to F_+$ such that $\Phi(x) = \Phi(-x) \ \forall x \in E, \forall \Phi \in \mathcal{S}$.

Corollary 2. *Let $E, F, \mathcal{S}$ be as above. Assume that Φ is continuous $\forall \Phi \in \mathcal{S}$ and $\sup_{\Phi \in \mathcal{S}} \|\Phi(x)\|_F < \infty \ \forall x \in E$. Then there exists a convex closed neighborhood U of 0_E such that $\bigcup_{\Phi \in \mathcal{S}} \Phi(U) \subset B_{1,F} := B_1(0_F)$, where $B_1(0_F)$ is the closed unit ball centered at the origin of the space F.*

The poof follows the ideas from that of Theorem 1, also applying Baire's theorem.

Remark 1. Under previous conditions, assuming that F is a normed vector lattice (the norm on F is solid and the lattice operations are continuous), Corollary 2 says, in particular, that

$$x_1 - x_2 \in U \Rightarrow |\Phi(x_1) - \Phi(x_2)| \leq \Phi(x_1 - x_2) \in B_{1,F} \Rightarrow$$

$$\Phi(x_1) - \Phi(x_2) \in B_{1,F} \quad \forall \Phi \in \mathcal{S}$$

It results that $\mathcal{S}$ is equicontinuous.

Example 1. Using the above notations, let $\mathcal{L}$ be a family of linear continuous operators from E to F such that $\sup_{T \in \mathcal{L}} \|T(x)\|_F < \infty$ $\forall x \in E$. Define $\Phi(x) = \Phi_T(x) := |T(x)|, x \in E, T \in \mathcal{L}$. Then the family $\mathcal{S} = \{\Phi_T\}_{T \in \mathcal{L}}$ verifies the condition $\sup_{T \in \mathcal{L}} \|\Phi_T(x)\|_F < \infty$ $\forall x \in E$.

Remark 2. In particular, Theorem 1 holds true when E is a Banach space, F is a normed vector lattice and the other conditions of Theorem 1 are accomplished. It is possible that a similar result be true for more general spaces E (involving the notion of a barreled TVS). However, only for a few spaces it can be easily proved that they are barreled spaces, without using Baire theorem. On the other side, for applications the most important spaces are Banach spaces, especially Banach lattices.

Theorem 2. *Let E be a Banach space, F an order complete normed vector lattice with strong order unit u_0, such that $B_{1,F} = [-u_0, u_0]$. Let $\mathcal{S}$ be a class of sublinear operators with the properties mentioned in Corollary 2. Additionally assume that $\Phi(x) = \Phi(-x)$ $\forall x \in E$. Then the relation*

$$\widetilde{\Phi}(x) = \sup_{\Phi \in \mathcal{S}} \Phi(x) \quad \forall x \in E$$

defines a sublinear Lipschitz operator $\widetilde{\Phi}$, such that $\widetilde{\Phi}(x) = \widetilde{\Phi}(-x) \in F_+$ $\forall x \in E$.
Proof. Application of Corollary 2 leads to the existence of a closed ball of sufficiently small radius $r > 0$ such that

$$\|x\|_E \leq r \Rightarrow \Phi(x) \in B_{1,F} = [-u_0, u_0] \quad \forall \Phi \in \mathcal{S}$$

It results

$$\Phi\left(r\frac{x}{\|x\|_E}\right) \leq u_0 \quad \Leftrightarrow \quad \Phi(x) \leq \frac{\|x\|_E}{r} u_0 \quad \forall x \in E\backslash\{0_E\}, \forall \Phi \in \mathcal{S} \tag{1}$$

Thus, according to (1), for any fixed $x \in E$ the set $\{\Phi(x); \Phi \in \mathcal{S}\}$ is bounded from above in F. Thanks to the hypothesis on order completeness of F, there exists

$$\widetilde{\Phi}(x) := \sup_{\varphi \in \mathcal{S}} \Phi(x) \leq \frac{\|x\|_E}{r} u_0 \quad \forall x \in E \tag{2}$$

It is easy to see that $\widetilde{\Phi}$ is sublinear and has the property $\widetilde{\Phi}(x) = \widetilde{\Phi}(-x) \in F_+$ $\forall x \in E$. Next we prove the Lipschitz property of $\widetilde{\Phi}$. To do this, one uses the subadditivity property of $\widetilde{\Phi}$, the fact that the norm of F is monotone on F_+, as well as relation (2). Namely, the following implications hold

$$x_1, x_2 \in E, \left|\widetilde{\Phi}(x_1) - \widetilde{\Phi}(x_2)\right| \leq \widetilde{\Phi}(x_1 - x_2) \Rightarrow$$

$$\left\|\widetilde{\Phi}(x_1) - \widetilde{\Phi}(x_2)\right\|_F \leq \left\|\widetilde{\Phi}(x_1 - x_2)\right\|_F \leq \left\|\frac{\|x_1 - x_2\|_E}{r} u_0\right\|_F = \frac{\|x_1 - x_2\|_E}{r}$$

Hence $\widetilde{\Phi}$ is a Lipschitz mapping from E to F_+. This concludes the proof. $\qquad\square$

Remark 3. Under the hypothesis of Theorem 2, each element of $\Phi \in \mathcal{S}$ is a Lipschitz operator, with the same Lipschitz constant $1/r$.

Remark 4. It seems that topological completeness of F is not necessary for the above results. However, the usual concrete spaces verifying the hypothesis of Theorem 2 are Banach spaces.

Remark 5. *The* set $\mathcal{C}$ of all continuous sublinear operators Φ from E to F_+, such that $\Phi(x) = \Phi(-x) \ \forall x \in E, \forall \Phi \in \mathcal{C}, \ \sup\limits_{\varphi \in \mathcal{C}} \|\varphi(x)\|_F < \infty \ \forall x \in E$, is a convex cone. With the notations and under the assumptions of Theorem 2, the subset of all $\Phi \in \mathcal{C}$ formed by all elements of $\mathcal{C}$ with the property $\varphi\left(B_{1,E}\right) \subset B_{1,F}$, is convex, and its elements are the nonexpansive operators from $\mathcal{C}$. If r from the proof of Theorem 2 is strictly greater than 1, then the elements of $\mathcal{S}$, (as well as the operator $\widetilde{\Phi}$), are contractions.

Remark 6. An arbitrary sublinear operator $\varphi \colon E \to F_+$ is a Lipschitz operator if and only if Φ is continuous at 0_E.

Corollary 3. *Let E, F be as in Theorem 2, $\mathcal{S} = \{\Phi_n; n \in N\}$ a countable set of sublinear continuous operators from E to F, such that $\Phi_n(x) = \Phi_n(-x) \ \forall x \in E, \forall n \in \mathbb{N}$, and $\sup\limits_{n \in \mathbb{N}} \|\Phi_n(x)\|_F < \infty \ \forall x \in E$. Then the relation*

$$\widetilde{\Phi}(x) = \sup\limits_{n \in \mathbb{N}} \Phi_n(x) \quad \forall x \in E$$

defines a sublinear Lipschitz operator $\widetilde{\Phi} \colon E \to F_+$, such that $\widetilde{\Phi}(x) = \widetilde{\Phi}(-x) \ \forall x \in E$.

Corollary 4. *Let E, F be as in Theorem 2, $\mathcal{T} = \{\varphi_n; n \in N, n \geq 1\}$ a countable set of sublinear continuous operators from E to F_+, such that $\Phi_n(x) = \Phi_n(-x) \ \forall x \in E, \forall n \in \{1, 2, \dots\}$, and*

$$\sup\limits_{\substack{n \in \mathbb{N}, \\ n \geq 1}} \left\|\sum_{k=1}^{n} \Phi_k(x)\right\|_Y < \infty \ \forall x \in E$$

Then the relations

$$\widetilde{\Phi}(x) = \sup_{\substack{n \in \mathbb{N}, \\ n \geq 1}} \left(\sum_{k=1}^{n} \Phi_k(x) \right) \quad \forall x \in E$$

define a sublinear Lipschitz operator $\widetilde{\Phi} \colon E \to F_+$, *such that* $\widetilde{\Phi}(x) = \widetilde{\Phi}(-x) \ \forall x \in E$.

Example 2. Let K be a compact Hausdorff topological space, μ a regular Borel probability measure on K, $E := C(K)$ the space of all real valued, continuous functions on K, $F := l_\infty$ the space of all bounded sequences of real numbers. The norm $\|\cdot\|_{sup}$ on the space E is the sup-norm and the norm on $\|\cdot\|_F$ is the usual norm $\|\cdot\|_F = \|\cdot\|_\infty$, $\|(x_n)_{n \geq 1}\|_\infty = \sup_{n \geq 1} |x_n|$. The space $F = l_\infty$ verifies the hypothesis of Theorem 2, since it is an order complete normed vector lattice, the appropriate strong order unit being the sequence u_0 which has all the terms equal to 1. Define the scalar valued norms on E

$$N_k(f) := \left(\int_K |f|^k \, d\mu \right)^{1/k}, f \in E, k \in \mathbb{N}, k \geq 1,$$

and the finite dimensional vector-valued norms on E

$$\varphi_n(f) := \|f\|_n \colon E \to F, \|f\|_n := \left(N_1(f), 2^{1/2} N_2(f), \ldots, n^{1/n} N_n(f), 0, \ldots, 0, \ldots \right),$$

$$n \in \mathbb{N}, n \geq 1, f \in E$$

$$N_k(f) \leq \|f\|_{sup} \left(\mu(K) \right)^{1/k} = \|f\|_{sup}, N_k(\P) = 1 \Rightarrow \sup_{\|f\|_{sup}=1} N_k(f) = 1,$$

$$k \in \{1, 2, \ldots\}, f \in E$$

Consider the elementary function

$$t \to g(t) := \ln(t)/t, t \in [1, \infty),$$

which is increasing on $[1, e]$ and decreasing on the interval $[e, \infty)$. This function has a global maximum point at $t_0 = e \in (2,3)$. It results that the function $h \colon (1, \infty) \to (0, \infty)$, $h(t) := t^{1/t} = e^{\ln(t)/t}$ has the same monotonicity properties, hence

$$\max_{1 \leq k \leq n} k^{1/k} \leq \max\{2^{1/2}, 3^{1/3}\} = 3^{1/3} \ \forall n \in \{1, 2, \ldots\}$$

Thus, we obtain

$$f \in X \Rightarrow \Phi_n(f) = \|f\|_n \leq \max_{1 \leq k \leq n} k^{1/k} \|f\|_{sup} u_0 \leq 3^{1/3} \|f\|_{sup} u_0 \ \forall n \in \{1, 2, \ldots\} \Rightarrow$$

$$\tilde{\Phi}(f) = \sup_{n \in \mathbb{N}} S_n(f) = \left(n^{1/n} N_n(f)\right)_{n \geq 1} \leq 3^{1/3} \cdot \|f\|_{sup} u_0, u_0 = (1, \ldots, 1, \ldots), f \in E$$

where $\tilde{\Phi}$ is the sublinear operator from Corollary 3. Observe that $\tilde{\Phi}$ has as Lipschitz constant $3^{1/3} > 1$. Next we apply the same method, replacing $n^{1/n}$ by

$$n^{-1/n} = exp(-\ln(n)/n) \leq 1 \ \forall n \in \{1, 2, \ldots\}.$$

In this case, the above evaluations change into the following ones

$$\tilde{\Phi}(f) = \sup_{n \in \mathbb{N}} \Phi_n(f) = \left(n^{-1/n} N_n(f)\right)_{n \geq 1} \leq \|f\|_{sup} u_0 \ \forall f \in E \Rightarrow$$

$$\left|\tilde{\Phi}(f) - \tilde{\Phi}(g)\right| \leq \tilde{\Phi}(f - g) \leq \|f - g\|_{sup} u_0 \Rightarrow$$

$$\left\|\tilde{\Phi}(f) - \tilde{\Phi}(g)\right\|_F \leq \left\|\tilde{\Phi}(f - g)\right\| \leq \|f - g\|_{sup} \ \forall f, g \in E$$

To conclude, in this case $\tilde{\Phi}$ is a nonexpansive vector valued norm from E to F. To obtain contractions $\tilde{\Phi}$, consider

$$(c_n)_{n \geq 1} \in Y = l_\infty, 0 \leq c_n \leq q < 1 \ \forall n \geq 1,$$

$$\tilde{\Phi}_n(f) = (c_1 N_1(f), \ldots, c_n N_n(f), 0, \ldots, 0, \ldots),$$

$$\tilde{\Phi}(f) = \sup_{n \geq 1} \tilde{\Phi}_n(f) = \left(c_n N_n(f)\right)_{n \geq 1} \leq q \|f\|_{sup} u_0 \ \forall f \in E \Rightarrow$$

$$\left|\tilde{\Phi}(f) - \tilde{\Phi}(g)\right|_F \leq \tilde{\varphi}(f - g) \leq q \|f - g\|_{sup} u_0 \Rightarrow$$

$$\left\|\tilde{\Phi}(f) - \tilde{\Phi}(g)\right\|_F \leq q \|f - g\|_{sup} \|u_0\|_L = q \|f - g\|_{sup} \ \forall f, g \in E$$

Thus $\tilde{\Phi} \colon E \to F_+$ is a contraction vector-valued norm, of contraction constant q, and the best value for q is $q = \sup_{n \geq 1} c_n$. In particular, if $0 \leq \inf_{n \geq 1} c_n \leq \sup_{n \geq 1} c_n = 1/2 \ \forall n \geq 1$, then $\tilde{\Phi}$ is a contraction operator, of contraction constant $q = 1/2$. In this example, the operators Φ_n mentioned in Corollary 4 stand for $\Phi_n(f) = (0, \ldots, 0, c_n N_n(f), 0, 0, \ldots)$, where $c_n N_n(f)$ is the $n - th$ coordinate of the vector $\Phi_n(f) \in F_+$.

3. Relationship between linear and sublinear continuous operators

In Theorem 3 proved below, we point out some remarks related on fixed points (if any) of some special sublinear operators.

Theorem 3. *Let E be an order complete normed vector lattice, $\Phi \colon E \to E_+$ a sublinear operator which is continuous at 0_E and verifies the conditions $\Phi(x) = \Phi(-x) \ \forall x \in E$. Assume that there exists a fixed point $x_0 \neq 0_E$ of Φ. Denote by $\mathcal{L}_0$ the set of all linear operators from E to E, with the following properties $T(x_0) = x_0, T(x) \leq \Phi(x) \ \forall x \in E$.*

Define $\quad \Phi_T(x) := |T(x)|, x \in E, T \in \mathcal{L}_0.$ *Then* $\quad S := \{\Phi_T; T \in \mathcal{L}_0\}$ *is a nonempty equicontinuous family of sublinear operators, for which the following relations hold true*

$$\Phi_T(x_0) = x_0, \quad \Phi_T(x) \leq \varphi(x) \ \forall x \in E, \quad \forall T \in \mathcal{L}_0 \tag{3}$$

Proof. According to Remark 7, $\mathcal{L}_0$ is nonempty, hence so is S. We next prove that $\mathcal{L}_0$ is equicontinuous. Continuity of Φ at 0_E leads to the existence of a sufficiently small $r > 0$ such that $\|x\|_E \leq r \Rightarrow \|\Phi(x)\|_E \leq 1$. This further yields $\left\| \Phi \left(r \frac{x}{\|x\|} \right) \right\|_X \leq 1 \ \forall x \in E, x \neq 0_E$, which can be rewritten as $\|\Phi(x)\|_E \leq \frac{1}{r} \|x\| := c\|x\|, \forall x \in E$. On the other side,

$$T(x) \leq \Phi(x) \ \forall x \in E \Rightarrow T(x) \leq \Phi(x), -T(x) = T(-x) \leq \Phi(-x) = \Phi(x) \ \forall x \in E \Rightarrow$$

$$\Phi_T(x) = |T(x)| \leq \Phi(x) \ \forall x \in E \tag{4}$$

(recall that the range of Φ is contained in E_+). Since the norm on E is solid, we infer that $\|T(x)\|_E \leq \|\Phi(x)\|_E \leq c\|x\|_E \ \forall x \in E, c := \frac{1}{r}, \forall T \in \mathcal{L}_0$. It results $\|T\|_b \leq c \ \forall T \in \mathcal{L}_0$, so that $\mathcal{L}_0$ is equicontinuous. It is also a convex subset of the space $B(E)$ of all bounded linear operators applying E to itself. Here $\|\cdot\|_b$ denotes the usual operatorial norm on $B(E)$. To finish the proof, we observe that $\Phi(x_0) \in E_+$ implies

$$\Phi_T(x_0) = |T(x_0)| = |x_0| = |\Phi(x_0)| = \Phi(x_0) = x_0 \ \forall T \in \mathcal{L}_0$$

Thus the first property (3) is verified. The second property (3) has already been proved by (4). Using (4) once more and the inequality $\|\Phi(x)\|_E \leq c\|x\|_E \ \forall x \in E$, we derive

$$\|\Phi_T(x)\|_E \leq \|\Phi(x)\|_E \leq c\|x\|_E \ \forall x \in E \tag{5}$$

Now the subadditivity of each Φ_T and (5) lead to

$$|\Phi_T(x_1) - \Phi_T(x_2)| \leq \Phi_T(x_1 - x_2) \Rightarrow$$

$$\|\Phi_T(x_1) - \Phi_T(x_2)\|_X \leq \|\Phi_T(x_1 - x_2)\| \leq c\|x_1 - x_2\|_E, \forall x_1, x_2 \in E, \forall T \in \mathcal{L}_0$$

This proves the equicontinuity of S and concludes the proof. $\qquad \square$

Proposition 1. *Let E be a reflexive Banach space endowed with a linear order relation defined by a closed positive cone E_+, and $B \subset E_+$ a convex bounded closed subset such that any $x \in E_+ \backslash \{0_E\}$ can be represented uniquely as $x = \rho b$ for some $\rho \in (0, \infty)$ and $b \in B$ (B is a base for E_+). Assume that F is a topological vector space endowed with an order relation defined by a closed positive cone F_+, $\Phi: E \to F$ is a bounded sublinear operator, and $T \in B(E, F)$ is a bounded linear operator such that*

$$\Phi(e) \leq T(e) \quad \forall e \in Ex(B)$$

Then $\Phi(x) \leq T(x)$ $\forall x \in E_+$.

Proof. Clearly, B is weakly compact and convex, so that $B = cl\left(co(Ex(B))\right)$. The topological closure of the convex set $co(Ex(B))$ in the weak topology equals its topological closure in the norm topology on E. Let

$$x_n = \sum_{j=1}^{n} \alpha_j e_j \in co(Ex(B)), \alpha_j \in [0, \infty), \sum_{j=1}^{n} \alpha_j = 1, e_j \in Ex(B), j = 1, \dots, n$$

Then by the assumptions of the statement we derive

$$\Phi(x_n) \leq \sum_{j=1}^{n} \alpha_j \Phi(e_j) \leq \sum_{j=1}^{n} \alpha_j T(e_j) = T\left(\sum_{j=1}^{n} \alpha_j e_j\right) = T(x_n), n \in \mathbb{N}, n \geq 1$$

For $b \in cl\left(co(Ex(B))\right) = B, b = \lim_n x_n, \ x_n \in co(Ex(B))$ for all $n \geq 1$, the continuity of Φ, T, as well as the hypothesis that the positive cone of F is closed, lead to

$$\Phi(b) = \lim_n \Phi(x_n) \leq \lim_n T(x_n) = T(b)$$

Now let $x \in E_+ \backslash \{0_E\}, x = \rho b, \rho > 0, b \in B$. Then

$$\Phi(x) = \Phi(\rho b) = \rho \Phi(b) \leq \rho T(b) = T(\rho b) = T(x)$$

This concludes the proof. $\qquad\qquad\qquad\qquad\qquad\qquad\qquad\qquad\qquad\quad \square$

Proposition 2. *Under the hypothesis of Proposition 1, additionally assume that E, F are normed vector lattices (their norms are solid), and Φ is isotone. Then $\|\Phi\| \leq \|T\|$.*

Proof. According to Proposition 1, we have already seen that $\Phi(w) \leq T(w)$ for all $w \in E_+$. Using this and monotonicity of Φ, we derive

$$(\Phi(x) \leq \Phi(|x|) \leq T(|x|), -\Phi(x) \leq \Phi(-x) \leq \Phi(|x|) \leq T(|x|)) \Rightarrow$$

$$|\Phi(x)| \leq T(|x|) \Rightarrow \|\Phi(x)\| \leq \|T\|\|x\| \ \forall x \in E \Rightarrow \|\Phi\| \leq \|T\|$$

This concludes the proof. $\qquad\qquad\qquad\qquad\qquad\qquad\qquad\qquad\qquad\quad \square$

CHAPTER 6

MOMENT PROBLEM, FINITE-SIMPLICIAL SETS AND RELATED SANDWICH RESULTS

1. Introduction

We start by completing a result of Chapter 1, first published in [37]. This complete statement (see Theorem 1.1 below) was applied in [2] to deduce a sandwich type result of intercalating an affine functional $h: X \to \mathbb{R}$ between a convex functional $f: X \to \mathbb{R}$ and a concave functional $g: X \to \mathbb{R}, f \leq g$, where X is a finite simplicial subset of a real vector space E. Here the novelty is that a finite simplicial set X may be unbounded in any locally convex topology on E (see Section 2 below and the paper [2]). In the present chapter, we prove a topological version of the sandwich result mentioned above (see Theorem 2.3 below). Some other related results are recalled or respectively proved. It is worth noticing that, conversely with respect to direct Hahn-Banach theorems, here the dominating functional g is concave, while the minorating functional f is convex. A previous such sandwich result for real functions defined on a Choquet simplex X is recalled in [56] and in Section 2 below.

Theorem 1.1. *Let E be an ordered vector space, F an order complete vector lattice, $\{\varphi_j\}_{j \in J} \subset E, \{y_j\}_{j \in J} \subset F$ given arbitrary families, $T_1, T_2 \in L(E, F)$ two linear operators. The following statements are equivalent*

(c) *there is a linear operator $T \in L(E, F)$ such that*

$$T_1(x) \leq T(x) \leq T_2(x) \ \forall x \in E_+, T(\varphi_j) = y_j \ \forall j \in J;$$

(d) *for any finite subset $J_0 \subset J$ and any $\{\lambda_j\}_{j \in J_0} \subset R$, the following implication holds true*

$$\left(\sum_{j \in J_0} \lambda_j \varphi_j = \psi_2 - \psi_1, \psi_1, \psi_2 \in E_+ \right) \Rightarrow \sum_{j \in J_0} \lambda_j y_j \leq T_2(\psi_2) - T_1(\psi_1).$$

If E is a vector lattice, then assertions (a) and (b) are equivalent to (c), where

(c) $T_1(w) \leq T_2(w)$ *for all $w \in E_+$ and for any finite subset $J_0 \subset J$ and $\forall \{\lambda_j; j \in J_0\} \subset \mathbb{R}$, we have*

$$\sum_{j \in J_0} \lambda_j y_j \leq T_2 \left(\left(\sum_{j \in J_0} \lambda_j \varphi_j \right)^+ \right) - T_1 \left(\left(\sum_{j \in J_0} \lambda_j \varphi_j \right)^- \right)$$

71

The rest of this chapter is organized as follows. In Section 2, topological versions for known sandwich results as those mentioned above are proved. Related remarks are discussed. Section 3 is devoted to extending inequalities from smaller subsets to the entire positive cone E_+, by means of Krein-Milman and respectively Carathéodory's theorem. Section 4 (the last one) refers to a direct sharp proof of a generalization of Hahn-Banach theorem, motivated by the results in this book, and, on the other hand, by recent results published in [31]. Additional remarks on the case when the dominating operator for the solution is sublinear and continuous are emphasized.

2. Finite simplicial sets and related sandwich theorems

We start by a preliminary lemma. Recall that a base B of a convex cone C contained in a vector space E is a set of the form $B = C \cap H$, where H is a hyperplane which misses the origin, defined by a strictly positive linear functional

$$\varphi: E \to \mathbb{R} \; (\varphi(x) > 0 \text{ for all } x \in C, x \neq 0)$$

If such a functional does exist, then choose any real $\beta > 0$ and define

$$H = \{x \in E; \varphi(x) = \beta\}$$

Then for any $x \in C, x \neq 0$, there exists a unique $b \in B = H \cap C$ and a unique real number

$$\alpha > 0 \text{ such that } x = \alpha b$$

Indeed, if $b = \beta \dfrac{x}{\varphi(x)}$, then $b \in B$ and

$$x = \frac{\varphi(x)}{\beta} b = \alpha b.$$

For the first result (Lemma 2.1), we will restrict ourselves to the Banach lattice setting, since the topological properties involved are important. In this respect, the hyperplane H and the positive cone E_+ are closed subsets (φ is continuous).

Lemma 2.1. *Let E, F be Banach lattices. Assume that the positive cone E_+ has a base*

$$B = H \cap E_+,$$

where H is a closed hyperplane missing the origin, as mentioned above. Let $p, -q: B \to F$ be convex continuous bounded operators on B and $\varphi: B \to F$ an affine continuous bounded operator. Define

$$\Phi, \Psi: E_+ \to F, \Phi(\lambda b) = \lambda p(b), \Psi(\lambda b) = \lambda q(b), b \in B, \lambda \in \mathbb{R}_+$$

Then Φ is a sublinear continuous extension of p, Ψ is supralinear, continuous and extends q, while

$$T(\lambda b) = \lambda \varphi(b)$$

defines an additive positively homogeneous operator on E_+, which can be further extended to a linear continuous operator T from E to F. In addition, we have

$$p \leq \varphi \leq q \text{ on } B \text{ if and only if } \Phi \leq T \leq \Psi \text{ on } E_+.$$

Proof. By definition, Φ, Ψ, T extend p, q, φ respectively, from B to E_+. To prove that Φ is subadditive, let $x_1 = \lambda_1 b_1, x_2 = \lambda_2 b_2$ be two elements in the positive cone E_+, where

$$b_j \in B, \lambda_j \in [0, \infty), j = 1,2, \lambda_1 + \lambda_2 > 0.$$

Then

$$\Phi(x_1 + x_2) = \Phi\left((\lambda_1 + \lambda_2)\left(\frac{\lambda_1}{\lambda_1 + \lambda_2}b_1 + \frac{\lambda_2}{\lambda_1 + \lambda_2}b_2\right)\right) :=$$

$$(\lambda_1 + \lambda_2)p\left(\frac{\lambda_1}{\lambda_1 + \lambda_2}b_1 + \frac{\lambda_2}{\lambda_1 + \lambda_2}b_2\right) \leq$$

$$(\lambda_1 + \lambda_2)\left(\frac{\lambda_1}{\lambda_1 + \lambda_2}p(b_1) + \frac{\lambda_2}{\lambda_1 + \lambda_2}p(b_2)\right) =$$

$$\lambda_1 p(b_1) + \lambda_2 p(b_2) = \Phi(x_1) + \Phi(x_2)$$

Next we prove that Φ is positively homogeneous: if $x = \lambda b, b \in B, \lambda \in (0, \infty), \alpha \in (0, \infty)$, then

$$\Phi(\alpha x) = \Phi\big((\alpha\lambda)b\big) = (\alpha\lambda)p(b) = \alpha\big(\lambda p(b)\big) = \alpha\Phi(x)$$

Thus Φ is sublinear, Ψ is supralinear and T *is* additive and positively homogeneous on the positive cone of E since φ is assumed to be simultaneous convex and concave on B. Extend T to the space

$$E = E_+ - E_+ \text{ by } T(x_1 - x_2) := T(x_1) - T(x_2), \text{ where } x_j \in E_+, j = 1,2$$

The definition makes sense, since it does not depend on decomposing x as a difference of two arbitrary elements from the positive cone E_+. Indeed, if $x_1 - x_2 = y_1 - y_2$, then

$$x_1 + y_2 = x_2 + y_1, x_j, y_j \in E_+, j = 1,2.$$

Now additivity of T on E_+ yields

$$T(x_1) + T(y_2) = T(x_2) + T(y_1) \Leftrightarrow T(x_1) - T(x_2) = T(y_1) - T(y_2)$$

For each nonnegative scalar λ, clearly it results

$$T(\lambda x) = T(\lambda x_1) - T(\lambda x_2) = \lambda\big(T(x_1) - T(x_2)\big) = \lambda T(x), x_j \in E_+, j = 1,2$$

On the other hand, according to definition from above, we infer that

$$T(-x) = T(-(x_1 - x_2)) = T(x_2) - T(x_1) = -\big(T(x_1) - T(x_2)\big) = -T(x)$$

If $\lambda < 0$, then $\lambda = -\rho$, where $\rho = -\lambda > 0$. From the preceding remark, we derive

$$T(\lambda x) = T(-\rho x) = \rho T(-x) = -\rho T(x) = \lambda T(x)$$

To finish the proof, we only have to prove the assertions on the continuity. If $(x_n)_{n \geq 1}$ is a sequence of elements from E_+, such that

$$x_n = \lambda_n b_n \to x = \lambda b, b_n, b \in B, \lambda_n, \lambda > 0,$$

then the following relations hold true

$$\varphi(x_n) = \lambda_n \varphi(b_n) = \lambda_n \beta \to \varphi(x) = \lambda \beta,$$

since φ is continuous, being a positive linear functional on the Banach lattice E. Thus $\lambda_n \to \lambda$. These relations lead to

$$b_n = \frac{1}{\lambda_n} x_n \to \frac{1}{\lambda} x = b$$

Due to the continuity of p on B, it results $p(b_n) \to p(b)$, which further implies

$$\Phi(x_n) = \lambda_n p(b_n) \to \lambda p(b) = \Phi(x)$$

A special case is $x = 0$, when

$$x_n = \lambda_n b_n \to 0 \Rightarrow \lambda_n \beta = \varphi(x_n) \to 0 \Rightarrow \lambda_n \to 0 \Rightarrow$$

$$\Phi(x_n) = \lambda_n p(b_n) \to 0,$$

since p is bounded on B by hypothesis. Thus the continuity of Φ, Ψ and T on E_+ is proved. We only have to show that the linear extension of T to the whole space E is continuous, where this extension has been denoted by T too. Clearly, the continuity at the origin will be sufficient. The following implications hold true

$$x_n \to 0 \Rightarrow x_n^+ \to 0 \Rightarrow T(x_n^+) \to 0$$

Similarly, $T(x_n^-) \to 0$, hence $\mathrm{T}(x_n) = T(x_n^+) - T(x_n^-) \to 0$. Finally, observe that

$$p \le h \le q \text{ on } B \Leftrightarrow \lambda p(b) = \Phi(\lambda b) \le \lambda h(b) = T(\lambda b) \le \lambda q(b) = \Psi(\lambda b), \lambda > 0, b \in B \Leftrightarrow$$

$$\Phi(x) \le T(x) \le \Psi(x), x \in E_+$$

The proof is complete. $\qquad\qquad\qquad\qquad\qquad\qquad\qquad\qquad\qquad\qquad\qquad$ $\square$

Next we recall the definition of a simplex in an infinite dimensional locally convex space E, and we emphasize one of its main properties. If C is a convex cone in E, C has a base B and the order relation defined by C is laticial on $E_1 = C - C$, then B is called a Choquet simplex. Usually, a simplex is assumed to be compact. For more information on simplexes see [56]. Now we recall the statement of *D.A. Edwards' separation theorem* (Theorem 16.7 [56]).

Lemma 2.2 (Edwards). *If f and $-g$ are convex upper semicontinuous real valued functions on a simplex B contained in a locally convex space, with $f \le g$, then there exists a continuous affine function h on B such that $f \le h \le g$.*

The next result is a topological version of Lemma 2.2.

Theorem 2.1. *Let E be a locally convex space, C a convex cone in E which has a simplex B as a base. Assume that the trace of the topology of E on $E_1 = C - C$ is locally solid with respect to the order relation defined by C. Let $\Phi: C \to \mathbb{R}$ be a continuous sublinear functional, $\Psi: C \to \mathbb{R}$ a continuous supralinear functional such that $\Phi(e) = \Psi(e)$ for all $e \in Ex(B)$. Then there exists a unique continuous linear functional $T: E_1 \to \mathbb{R}$ such that*

$$\Phi(x) \le T(x) \le \Psi(x) \; \forall x \in C \tag{2.1}$$

Proof. Let $f := \Phi|_B, g := \Psi|_B$. Then f is convex and continuous, g is concave and continuous, and $f \le g$ on B ($f - g$ is continuous and convex, vanishing on $Ex(B)$ by hypothesis; it results $(f - g)(x) \le 0$ for all $x \in co(Ex(B))$ and via continuity, $(f - g)(x) \le 0$ for all $x \in cl\left(co(Ex(B))\right) = B$, where the last equality is given by Krein-Milman theorem). Since B is a simplex, application of Lemma 2.2 leads to the existence of a continuous affine function $h: B \to \mathbb{R}$, such that $f \le h \le g$. According to the first part of the proof of Lemma 2.1 (which does not use any topological notions), h has a unique linear extension, say T, to $E_1 = C - C$, such that (2.1) holds true. The next step is to prove the continuity of T on E_1, which is equivalent to its continuity at the origin. Let $(x_\delta)_{\delta \in \Delta}$ be a generalized sequence in E_1, such that $x_\delta \to 0$. Consider the sequences $(x_\delta^+)_{\delta \in \Delta}, (x_\delta^-)_{\delta \in \Delta}$ in C. According to the assumptions on the topology on E_1, also using the continuity at the origin of Φ, Ψ, and (2.1) as well, we infer that

$$x_\delta \to 0 \Rightarrow (x_\delta^+ \to 0, x_\delta^- \to 0) \Rightarrow$$

$$(\Phi(x_\delta^+) \to 0, \Psi(x_\delta^+) \to 0, \Phi(x_\delta^+) \leq T(x_\delta^+) \leq \Psi(x_\delta^+)) \Rightarrow T(x_\delta^+) \to 0$$

Similarly, $T(x_\delta^-) \to 0$, so that

$$T(x_\delta) = T(x_\delta^+) - T(x_\delta^-) \to 0$$

Hence T is continuous. To finish the proof, we only have to show the uniqueness of T with the properties in the statement. Let T_1 be a linear continuous functional on E_1 such that (2.1) holds for T_1 instead of T. Let h_1 be the restriction of T_1 to the simplex B. Then h_1 is continuous and affine on B. Moreover, we have

$$h_1(e) = T_1(e) \in [\Phi(e), \Psi(e)] = \{T(e)\} = \{h(e)\} \, \forall e \in Ex(B) \Rightarrow$$
$$h_1(e) = h(e) \ \forall e \in Ex(B)$$

This further yield $h_1(x) = h(x)$ for all $x \in co\big(Ex(B)\big)$ (via the property of being affine for h, h_1). Now also using the continuity of the two involved functions, we derive

$$h_1(x) = h(x) \ \text{for all} \ x \in cl\left(co\big(Ex(B)\big)\right) = B,$$

where the last equality is a consequence of Krein-Milman theorem. Thus $T_1|_B = T|_B$, which means $T_1 = T$ on E_1. This ends the proof. $\qquad\qquad\square$

Remark 2.1. Let $E = l_1$ and F an arbitrary Banach space endowed with a linear order relation such that the positive cone F_+ is closed. Let $\{e_n\}_{n=1}^\infty$ be the canonical base of E and $\{y_n\}_{n=1}^\infty$ a bounded sequence in F. Assume that $\Phi, -\Psi \colon E_+ \to F$ are given sublinear continuous operators such that $\Phi(e_n) = \Psi(e_n)$ for all natural numbers $n \geq 1$. Then there exists a unique bounded linear operator $T \in \mathcal{L}(E, F)$ such that $\Phi \leq T|_{E_+} \leq \Psi$. Indeed, if we denote

$$y_n = \Phi(e_n), n \geq 1$$

and define

$$T(x) = \sum_{n=1}^{\infty} x_n y_n \, , x = (x_n)_{n \geq 1} \in E, \tag{2.2}$$

then the series on the right hand side is absolutely convergent in the Banach space F for each fixed element $x \in E$. Thus (2.2) defines a linear operator, which is continuous, with

$$\|T\| = \sup_{n \geq 1}\|y_n\| < \infty$$

From (2.2), also using the hypothesis, it results $T(e_n) = y_n = \Phi(e_n) = \Psi(e_n), n \geq 1$. This And continuity of all involved operators, as well as sublinearity of $\Phi, -\Psi$ lead to

$$\Phi(x) \leq T(x) \leq \Psi(x)$$

for all $x \in E_+$. It is possible that such examples could be adapted for more general spaces E.

In the end of this section, we recall that related sandwich theorems hold on the finitesimplicial sets, as discussed in [2]. A convex subset X of the vector space E is called finite-simplicial if for any finite dimensional convex compact subset $K \subset X$, there exists a finite dimensional simplex S such that $K \subset S \subset X$. The novelty here is that X is not supposed to be bounded in a locally convex topology on E. Here are a few examples.

1) In $\mathbb{R}^n, n \geq 2$, any convex cone X having a base that is a simplex (the corresponding order relation is laticial) is an unbounded finite simplicial set.
2) In $\mathbb{R}^n, n \geq 2$, for each $p \in (1, \infty)$, the convex cone X defined by

$$X = \left\{ (x_1, \dots, x_n); \ x_n \geq \left(\sum_{j=1}^{n-1} |x_j|^p \right)^{1/p} \right\}$$

has a compact base, but X is not finite-simplicial.
3) Let E be an arbitrary infinite or finite dimensional vector space (of dimension ≥ 2), $T: E \to \mathbb{R}$ a non-null linear functional and $\alpha \in \mathbb{R}$. Then the sets $X_1 = \{x; T(x) \geq \alpha\}, X_2 = \{x; T(x) \leq \alpha\}$ are finite-simplicial.
4) Let E, T be as in Example 3), α, β two real numbers such that $\alpha < \beta$. The set

$$X = \{x; \ \alpha \leq T(x) \leq \beta\}$$

is not finite-simplicial. From the last two examples, we easily infer that generally the intersection of two finite-simplicial sets is not finite-simplicial.

The following sandwich type result holds true (cf. [2]).

Theorem 2.2. *Let E be an arbitrary vector space, X a finite-simplicial subset, $f: X \to \mathbb{R}$ a convex functional, $g: X \to \mathbb{R}$ a concave functional such that $f \leq g$ on X. Then there exists an affine functional $h: X \to \mathbb{R}$ such that $f \leq h \leq g$.*

Here is a topological version of Theorem 2.2.

Theorem 2.3. *Let E be an ordered Banach space. Assume that E_+ is finite-simplicial and there exists $x_0 \in E_+$ such that $E_+ - x_0$ contains a balanced, absorbing, convex subset. Let*

$$f, -g: E_+ \to \mathbb{R}$$

be convex continuous functions such that $f \leq g$. Assume also that $f(0) = g(0) = 0$. Then here exists a continuous linear form $S: E \to \mathbb{R}$ such that $f \leq S \leq g$ on E_+.

Proof. According to Theorem 2.2, there exists an affine functional $h: E_+ \to \mathbb{R}$ such that

$$f \leq h \leq g \text{ on } E_+.$$

Using also the hypothesis, we infer that

$$0 = f(\mathbf{0}) \leq h(\mathbf{0}) \leq g(\mathbf{0}) = 0,$$

hence $h(\mathbf{0}) = 0$. On the other hand, since $E_+ - x_0$ is convex, absorbing and balanced, it is a neighborhood of $\mathbf{0}$ in the finest locally convex topology on E (see [59]). Recall that in this topology, any convex balanced absorbing subset is a neighborhood of $\mathbf{0}$ (this is the topology generated by the family of all seminorms on E). Hence there exists a convex open neighborhood V of $\mathbf{0}$ in this topology, such that

$$D = x_0 + V \subset E_+$$

In particular, the interior $in(E_+)$ of E_+ with respect to the finest locally convex topology on E is not empty. Denote this interior by C. Since h is convex and concave on the convex open subset C (in the finest locally convex topology on E) and $x_0 \in C$, the subdifferentials of h and $-h$ at x_0 are nonempty. This means that there exist linear functionals T and U on E such that

$$T(x) - T(x_0) \leq h(x) - h(x_0) \leq U(x) - U(x_0)$$

for all $x \in C$. Since clearly any linear form on E is continuous with respect to the finest locally convex topology, from the last two inequalities we infer that h is continuous at x_0. Repeating this argument for any $x_0 \in C$, we infer that h is continuous on C with respect to the topology under attention. As a consequence, the subsets

$$A = \{(x,r) \in C \times \mathbb{R}; h(x) < r\}, B = \{(x,r) \in C \times \mathbb{R}; h(x) > r\}$$

are open convex disjoint subsets in $C \times \mathbb{R}$ endowed with the product topology. By separation theorem, there exists a hyperplane $H \subset E \times \mathbb{R}$, separating (strictly) the subsets A and B. So, there is a not null linear form L on $E \times \mathbb{R}$ and $\alpha \in \mathbb{R}$ such that

$$H = \{(x,r) \in E \times \mathbb{R}; L(x,r) = \alpha\}, A \subset \{(x,r) \in C \times \mathbb{R}; L(x,r) < \alpha\},$$

$$B \subset \{(x,r) \in C \times \mathbb{R}; L(x,r) > \alpha\}$$

All the topological notions in the sequel refer to the finest locally convex topology, unless other specification is mentioned. Since L is linear, it is continuous on $E \times \mathbb{R}$. For $(x,r_1) \in A$, we have

$$L(x,0) + r_1 L(\mathbf{0},1) < \alpha,$$

while for any $(x, r_2) \in B$, it results

$$L(x, 0) + r_2 L(\mathbf{0}, 1) > \alpha$$

From these last two inequalities written for a $x \in C$ and $r_1 \neq r_2$, we infer that

$$L(x, 0) + r_1 L(\mathbf{0}, 1) < \alpha < L(x, 0) + r_2 L(\mathbf{0}, 1),$$

that implies $(r_2 - r_1)L(0,1) > 0$. In particular, it results $L(0, 1) \neq 0$. Going back to the definitions of the subsets A and B, observe that for any $x \in C$ and any $\varepsilon > 0$ the element $(x, h(x) + \varepsilon) \in A$, while $(x, h(x) - \varepsilon) \in B$. This remark yields

$$L(x, 0) + (h(x) + \epsilon)L(\mathbf{0}, 1) < \alpha, L(x, 0) + (h(x) - \epsilon)L(\mathbf{0}, 1) > \alpha$$

Passing to the limit as $\varepsilon \to 0$, one obtains

$$L(x, 0) + h(x)L(\mathbf{0}, 1) = \alpha$$

for all $x \in C$. This may be written as

$$h(x) = \frac{\alpha}{L(\mathbf{0}, 1)} - \frac{L(x, 0)}{L(\mathbf{0}, 1)}$$

Hence h is the restriction to $C = in(E_+)$ of the linear form S defined on E by $(x) = -\frac{L(x,0)}{L(0,1)}$, then adding the constant $\frac{\alpha}{L(0,1)}$. Recall now that we have already proved that $h(\mathbf{0}) = 0$, that yields $\alpha = 0$, so that h is the restriction to C of the linear form S. On the other hand, by hypothesis, f and g are assumed to be continuous with respect to the norm topology. In particular, they are continuous with respect to the finest locally convex topology and S has the latter property as well. We also have

$$f|_C \leq S|_C \leq g|_C$$

and $C = in(E_+)$ is dense in E_+. From the continuity of the three involved functionals, the inequalities written above are extended to the entire positive cone E_+. To finish the proof, we must show that S is continuous with respect to the norm topology, that is equivalent to its continuity at $\mathbf{0}$. From now on, only the norm topology is involved. Let $x_n \to \mathbf{0}$. As it is known, the ordered Banach space E can be renormed by means of an equivalent norm, such that it becomes a regularly ordered Banach space. In such a space for any sequence $x_n \to \mathbf{0}$, there exists sequences $(u_n)_n, (v_n)_n$, $u_n, v_n \in E_+, x_n = u_n - v_n$ for all $n \in \mathbb{N}$, such that $u_n \to \mathbf{0}, v_n \to \mathbf{0}$. It results

$$0 = f(\mathbf{0}) \leftarrow f(u_n) \leq S(u_n) \leq g(u_n) \to g(\mathbf{0}) = 0$$

which obviously imply $S(u_n) \to 0$. Similarly, $S(v_n) \to 0$, so that

$$S(x_n) = S(u_n) - S(v_n) \to 0$$

This concludes the proof. $\qquad\qquad\qquad\qquad\qquad\qquad\qquad\qquad\square$

3. Extending inequalities via Krein-Milman and Carathéodory's theorems

The next results extend an inequality occurring on a small set to a much larger subset.

Theorem 3.1. *Let E be a reflexive Banach lattice, F an order complete Banach lattice in which every topological bounded subset is order-bounded and $y_n \uparrow y$ implies $y_n \to y$, $\Phi: E_+ \to F$ a quasiconvex continuous positively homogeneous operator, $T \in B_+(E, F)$ a positive linear operator such that $\Phi(e) \leq T(e)$ for all extreme points e of the set $K :=$ $E_+ \cap B_{1,E}$. Then*

$$\Phi(x) \leq \|x\| \cdot \sup_{e \in Ex(K)} \Phi(e) \in F_+ \quad \forall x \in E_+$$

Proof. Recall that an operator Φ from a convex subset C of a vector space E to a vector lattice F is called quasiconvex if

$$\Phi\big((1-t)x_1 + tx_2\big) \leq sup\{\varphi(x_1), \varphi(x_2)\}, \forall t \in [0,1], \forall x_1, x_2 \in C$$

Following the proof (by induction) of Jensen's inequality for real quasiconvex functions, for any convex combination $\sum_{j=1}^{n} \alpha_j x_j$, $x_j \in C, j = 1, \ldots, n$, a quasiconvex operator Φ verifies

$$\Phi\left(\sum_{j=1}^{n} \alpha_j x_j\right) \leq sup\{\Phi(x_1), \ldots, \Phi(x_n)\} \tag{3.1}$$

See [30] for details, examples and exercises related to this important notion. The set $K :=$ $X_+ \cap B_{1,X}$ is convex, weakly compact and

$$K = cl\big(co(Ex(K))\big) \tag{3.2}$$

holds thanks to Krein-Milman theorem. Let

$$x_n = \sum_{j=1}^{n} \alpha_j e_j \in co\big(Ex(K)\big), e_j \in Ex(K), \alpha_j \in [0, \infty), j = 1, \ldots, n, \sum_{j=1}^{n} \alpha_j = 1$$

According to (3.1) and also using hypothesis, we infer that

$$\Phi(x_n) \leq sup\{\Phi(e_1), \ldots, \Phi(e_n)\} \leq sup\{T(e_1), \ldots, T(e_n)\} \tag{3.3}$$

On the other side, any positive linear operator from E to F is continuous, so that the image of the bounded set $K \subset E_+$ through the positive (bounded) linear operator T is topologically bounded, hence is o − bounded in F_+. Thus it results

$$T(K) \subset [0_F, y_0] \text{ for some } y_0 \in F_+$$

From this and also using (3.3), it results

$$\Phi(x_n) \leq y_0, x_n \in co\big(Ex(K)\big), n \in \mathbb{N}, n \geq 1$$

If $x = \lim_{n \to \infty} x_n \in cl\left(co\big(Ex(K)\big)\right) = K$, where $x_n \in co\big(Ex(K)\big)$ for all $n \geq 1$, then, thanks to the continuity of Φ, we are leaded to

$$\Phi(x) = \lim_{n \to \infty} \Phi(x_n) \leq \lim_{n \to \infty} \big(sup\{\Phi(e_j); j = 1, ..., n\}\big) =$$

$$\sup_{n \geq 1} \Phi(e_n) \leq \sup_{e \in Ex(K)} \Phi(e) \leq \sup_{e \in Ex(K)} T(e) \leq y_0, x \in K$$

Since Φ is positively homogeneous, application of this evaluation to $x/\|x\| \in K$, for all $x \in X_+, x \neq 0_X$, yields

$$\Phi(x) \leq \|x\| \cdot \sup_{e \in Ex(K)} \Phi(e) \leq \|x\| y_0, \quad x \in E_+$$

This concludes the proof. □

Theorem 3.2. *Let E be an order complete normed vector lattice, $K \subset E$ a finite dimensional compact subset, $(\Phi_n)_{n \geq 0}$ a sequence of continuous sublinear operators from E to E, such that for each $x \in E$, there exists $\widetilde{\Phi}(x) := \lim_{n \to \infty} \Phi_n(x) \in E$. Assume that for each $n \in \mathbb{N}$, there exists an affine operator T_n from X to X, such that $T_n(x) \leq \Phi_n(x) \; \forall x \in E$ and there exists*

$$\tilde{T}(e) := \lim_{n \to \infty} T_n(e) = e \; \forall e \in Ex(K).$$

Then $x \leq \widetilde{\Phi}(x) \; \forall x \in Cone(K)$, where $Cone(K)$ is the convex cone generated by $(co(K)) \cup \{0_E\}$.

Proof. It is known that for any finite dimensional compact K, its convex hull $co(K)$ is compact too (the proof of this assertion is based on Carathéodory's theorem, which leads to a way of expressing $co(K)$ as image of a compact (finite dimensional) subset through a continuous mapping). Let p be the dimension of the linear variety generated by K (and by $co(K)$) and $x \in co(K)$. Assume that $p \geq 2$. Due to Carathéodory's theorem, there exist at most $p + 1$ extreme points in the compact (convex) subset $co(K)$, say $e_1, ..., e_{p+1}$ and $\{\alpha_1, ..., \alpha_{p+1}\} \subset [0, \infty), \sum_{j=1}^{p+1} \alpha_j = 1$, such that $x = \sum_{j=1}^{p+1} \alpha_j e_j$. Also, it is known that any extreme point of $co(K)$ is (an extreme) point of K. From hypothesis we infer that

$$T_n(x) = \sum_{j=1}^{p+1} \alpha_j T_n(e_j) \to \sum_{j=1}^{p+1} \alpha_j \tilde{T}(e_j) = \sum_{j=1}^{p+1} \alpha_j e_j = x, \qquad n \to \infty$$

Thus, there exists $\tilde{T}(x) = x \ \forall x \in co(K)$. On the other side, the positive cone E_+ of the space E is closed and we have assumed that $\Phi_n(x) - T_n(x) \in E_+$ for all $x \in E$ and all $n \in \mathbb{N}$. Passing to the limit, one obtains

$$\tilde{\Phi}(x) - \tilde{T}(x) = \tilde{\Phi}(x) - x \in E_+ \quad \forall x \in co(K) \Leftrightarrow \tilde{\Phi}(x) \geq x \ \forall x \in co(K) \tag{3.4}$$

(Since $\tilde{T}$ is the pointwise limit of affine operators, it is affine on $co(K)$; a Hahn-Banach argument shows that it has an affine extension from the whole space E to E. We denote this extension by $\tilde{T}$ too). Recall that if 0_E is not an element of $co(K)$, then

$$C := Cone(K) = \{\alpha x; \ \alpha \in [0, \infty), x \in co(K)\} \tag{3.5}$$

It is easy to see that in this case: $C \cap (-C) = \{0_E\}$. Now (3.4) and (3.5) yield

$$\tilde{\Phi}(\alpha x) = \alpha \tilde{\Phi}(x) \geq \alpha x \ \forall \alpha \in [0, \infty), \forall x \in co(K) \Leftrightarrow \tilde{\Phi}(w) \geq w \ \forall w \in C$$

If $0_E \in \partial(co(K)) \backslash ri(co(K))$, then C could satisfy the condition $C \cap (-C) = \{0_E\}$, or $C \cap (-C)$ might be a nonzero vector subspace (here $ri(co(K))$ is the relative interior of $co(K)$). When $0_E \in ri(co(K))$, C is a $p -$ dimensional vector subspace of E. In both these last two cases, the conclusion of the theorem still holds true, following the same proof as in the first case. This concludes the proof. $\qquad \square$

Remark 3.1. Assume now that 0_E is not an element of $co(K)$. Then there exists a strictly positive linear continuous form T on E endowed with the order relation defined by C, such that $\|T\| = 1$, and a constant $\beta > 0$ with

$$\inf_{x \in co(K)} T(x) = \beta = T(e) \text{ for some } e \in Ex(co(K)).$$

Indeed, denote by $d_0 := d(0_E, co(K)) > 0, V := B_{d_0}(0_E) = \{x \in E; \|x\| < d_0\}$. Then V is a convex open neighborhood of the origin, which does not intersect $co(K)$. By geometric form of Hahn-Banach theorem, there exists a closed hyperplane separating V from $co(K)$, and not intersecting V i. e. there exists a linear continuous form T on E such that $0 < supT(V) \leq \beta \leq infT(co(K))$. In particular, $T(x) \geq \beta > 0 \ \forall x \in co(K) \Rightarrow T(w) > 0 \ \forall w \in C \backslash \{0_E\}$. Whence T is strictly positive and for any $\gamma > 0$, the set $B = \{x \in C; T(x) = \gamma\}$ is a compact base for C. Now scaling T by a positive scalar we obtain a new strictly continuous positive form (which we also denote by T) with the special property

$$0 < \sup T(V) = d_0 = \inf T\big(co(K)\big), T(x) < d_0 \ \forall x \in V$$

It results

$$T\left(d_0 \frac{x}{\|x\| + \varepsilon}\right) < d_0 \ \forall \varepsilon > 0, \forall x \in X \Rightarrow |T(x)| \le \|x\|, \forall x \in E \Rightarrow \|T\| \le 1$$

Let $x_\star \in co(K) \cap (\partial V)$ be such that $T(x_\star) = d_0$. Then $\|x_\star\| = d_0$ and $T\left(\frac{x_\star}{\|x_\star\|}\right) = 1$. Thus $\|T\| = 1$. Consider the hyperplane $H := \{x; T(x) = d_0\}$ and the base $B = H \cap C$. Then the distance

$$d(0_E, H) = \frac{|T(0) - d_0|}{\|T\|} = d_0 = d\big(0_E, co(K)\big),$$

as expected (H and B are separating $cl(V)$ and $co(K)$, but they are "tangent" to both these closed convex subsets). If E is a real Hilbert space and T has the properties from above, then $x_\star$ is the orthogonal (or metric) projection of 0_E to $co(K)$. Consequently,

$$\|x_\star\| = d\big(0_E, co(K)\big) = d_0, \qquad x_\star \perp H$$

Having in mind the idea of the proof of Riesz representation theorem for linear continuous forms on a Hilbert space, it results that T is represented by a vector which is collinear to $x_\star$. Since $\|T\| = 1$, we have to normalize $x_\star$. It results

$$T(x) = <\frac{x_\star}{\|x_\star\|}, x> = <\frac{x_\star}{d_0}, x> \quad \forall x \in E$$

4. A direct proof for a generalization of Hahn-Banach theorem and its motivation

The following theorem has been recently applied in [31] to a characterization of isotonicity of a continuous convex operator over a convex cone in terms of its subgradients. It can be obtained from more general results published in [34], [36]. The proof of these very general results is quite long and technical. Therefore, it is preferable to have a direct proof, using only Zorn lemma and appropriate inequalities.

Theorem 4.1. (Theorem 3 [31]). *Let E be an ordered vector space, F an order complete vector space, $H \subset E$ a vector subspace, $T_1: H \to F$ a linear operator, $\Phi: E_+ \to F$ a convex operator. The following statements are equivalent*

(a) *there exists a positive linear extension $T: E \to F$ of T_1 such that $T|_{E_+} \le \Phi$;*
(b) *we have $T_1(h) \le \Phi(x)$ for all $(h, x) \in H \times E_+$ such that $h \le x$.*

Proof. The implication $(a) \Rightarrow (b)$ is obvious; indeed, we have

$$T_1(h) = T(h) \le T(x) \le \Phi(x)$$

$(h, x) \in H \times E_+$ such that $h \leq x$, thanks to the positivity of T, also using the property $T(x) \leq \Phi(x)$ for all $x \in E_+$. To prove the converse, let Φ be an arbitrary convex operator over E_+, verifying the conditions mentioned at (b). We are going to apply Zorn lemma to the set $\mathcal{L}$ of all pairs (K, T_K), where K is a vector subspace of $E, H \subset K, T_K : K \to F$ is a linear operator such that $T_K|_H = T_1$ and $T_K(h) \leq \Phi(x)$ for all $h \in K$ and $x \in E_+$ such that $h \leq x$. The set $\mathcal{L}$ contains the pair (H, T_1) and is inductively ordered by the order relation

$$(K, T_K) \ll (L, T_L) \Leftrightarrow K \subset L, T_L|_K = T_K$$

According to Zorn's Lemma, there exists a maximal pair $\left(H_M, T_{H_M} \right) \in \mathcal{L}$. Our aim is to prove that $H_M = E$. Assuming this is done, and taking $T := T_{H_M}$, we have $T(h) \leq \Phi(x)$ for all $h \in H$ and $x \in E_+$ such that $h \leq x$. Application of this inequality for $h = -ny, y \in E_+, n \in \mathbb{N}, x = 0$, yields

$$nT(-y) = T(-ny) \leq \Phi(\mathbf{0}), n \in \mathbb{N}$$

Since any order complete vector space is Archimedean, it results $T(-y) \leq \mathbf{0}, y \in E_+$, that is the positivity of T. Also, taking $h = x \in E_+$, one obtains $T(x) \leq \Phi(x)$ and T will be the expected positive extension of $T_1, T|_{E_+} \leq \Phi$, and this will end the proof. Assuming that $H_M \neq E$, we can choose $v_0 \in E \backslash H_M$ and define $H_0 = H_M \oplus \mathbb{R}v_0, T_0 : H_0 \to F$,

$$T_0(h + rv_0) = T_{H_M}(h) + ry_0 , h \in H_M, r \in \mathbb{R}$$

where $y_0 \in F$ will be chosen such that $\left(H_M, T_{H_M} \right) \ll (H_0, T_0)$ and $(H_0, T_0) \in \mathcal{L}$. This will contradict the maximality of $\left(H_M, T_{H_M} \right)$ in $\mathcal{L}$. Thus $H_M = E$. To prove that $(H_0, T_0) \in \mathcal{L}$ for suitable $y_0 \in F$, we have to show that

$$h \in H_M, r \in \mathbb{R}, h + rv_0 \leq x \in E_+ \Rightarrow T_{H_M}(h) + ry_0 \leq \Phi(x)$$

For $r = \alpha > 0$, multiplying α^{-1} the relation $h_1 + \alpha v_0 \leq x_1 \in E_+$ the above implication becomes

$$h_1 + \alpha v_0 \leq x_1 \in E_+ \Rightarrow y_0 \leq \alpha^{-1} \left(\Phi(x_1) - T_{H_M}(h_1) \right), \alpha > 0,$$

and, respectively,

$$h_2 + \beta v_0 \leq x_2 \in E_+ \Rightarrow y_0 \geq \beta^{-1} \left(\Phi(x_2) - T_{H_M}(h_2) \right), \beta < 0$$

To have both conditions on y_0 verified, according to order completeness of F, it is necessary and sufficient to prove that

$$\beta^{-1} \left(\Phi(x_2) - T_{H_M}(h_2) \right) \leq \alpha^{-1} \left(\Phi(x_1) - T_{H_M}(h_1) \right)$$

The last inequality may be written as

$$T_{H_M}(\alpha^{-1} h_1 - \beta^{-1} h_2) \leq \alpha^{-1} \Phi(x_1) - \beta^{-1} \Phi(x_2)$$

To prove this last inequality, we eliminate v_0 by adding the previous inequalities, multiplied by $\alpha^{-1} > 0$, respectively by $-\beta^{-1} > 0$, as follows

$$(\alpha^{-1}h_1 + v_0 \leq \alpha^{-1}x_1, -\beta^{-1}h_2 - v_0 \leq -\beta^{-1}x_2) \Rightarrow \alpha^{-1}h_1 - \beta^{-1}h_2 \leq \alpha^{-1}x_1 - \beta^{-1}x_2 \in E_+,$$

where $h_j \in H_M, x_j \in E_+, j = 1,2, \alpha > 0, \beta < 0$. Since $(H_M, T_{H_M}) \in \mathcal{L}$ and Φ is convex, this further yields

$$\frac{1}{\alpha^{-1} - \beta^{-1}} T_{H_M}(\alpha^{-1}h_1 - \beta^{-1}h_2) = T_{H_M}\left(\frac{\alpha^{-1}}{\alpha^{-1} - \beta^{-1}}h_1 + \frac{-\beta^{-1}}{\alpha^{-1} - \beta^{-1}}h_2\right) \leq$$

$$\Phi\left(\frac{\alpha^{-1}}{\alpha^{-1} - \beta^{-1}}x_1 + \frac{-\beta^{-1}}{\alpha^{-1} - \beta^{-1}}x_2\right) \leq$$

$$\frac{\alpha^{-1}}{\alpha^{-1} - \beta^{-1}}\Phi(x_1) + \frac{-\beta^{-1}}{\alpha^{-1} - \beta^{-1}}\Phi(x_2)$$

Thus the expected inequality follows and the proof is complete. $\qquad\square$

Corollary 4.1. (Theorem 6 [31]). *Let E, F be ordered Banach spaces, F being order-complete, $P: E_+ \to F$ a continuous convex operator. Then P is isotone on E_+ if (and only if) for each $x_0 \in E_+$, there exists a positive continuous linear operator $T_{x_0}: E \to F$ such that*

$$P(w) - P(x_0) \geq T_{x_0}(w - x_0)$$

for all $w \geq x_0$.

Proof. Assume that P is as in the statement and is also isotone on E_+. We define $\Phi: E_+ \to F$,

$$\Phi(x) = P(x + x_0) - P(x_0), x \in E_+$$

The olperator Φ is convex, continuous and isotone, since so is P. In addition, $\Phi(0) = 0$. Application of Theorem 4.1 to $H = \{0\}, T_1 = 0 = \Phi(0) \leq \Phi(x)$ for all $x \in E_+$ (since Φ is isotone), yields the existence of a linear positive operator $T \in L_+(E, F)$ such that

$$T(x) \leq \Phi(x) = P(x + x_0) - P(x_0), x \in E_+ \Leftrightarrow$$

$$T(w - x_0) \leq P(w) - P(x_0), \qquad w := x + x_0 \geq x_0$$

According to Lemma 2 [31], positivity of T implies its continuity. Conversely, assume that the inequality in the statement holds true for each $x_0 \in E_+$ and any $w \geq x_0$. Let $0 \leq x \leq y$ in E. Application of the claimed inequality to $x_0 = x, w = y$ leads to

$$P(y) - P(x) \geq T_x(y - x) \geq 0,$$

since T_x is positive and $y - x \geq 0$. Thus $P(y) \geq P(x)$ and this ends the proof. $\quad\square$

Having in mind the same proof as that of Theorem 4.1, the following variant also holds true, according to the above mentioned proof.

Theorem 4.2. *Let E be an ordered vector space, F an order complete vector space, $H \subset E$ a vector subspace, $T_1 \colon H \to F$ a linear operator, $\Phi \colon E \to F$ a convex operator. The following statements are equivalent*

    (a) *there exists a positive linear extension $T \colon E \to F$ of T_1 such that $T \leq \Phi$ on E;*
    (b) *we have $T_1(h) \leq \Phi(x)$ for all $(h, x) \in H \times E$ such that $h \leq x$.*

Remark 4.1. Theorem 2.2 of Chapter 1 represents another way of writing Theorem 4.2, in terms of the abstract moment problem ($Span\{x_j; j \in J\}$ from Theorem 2.2 of Chapter 1 stands for H from Theorem 4.2). It results two motivations of proving results such as Theorems 4.1 and 4.2: (i) characterizing isotonicity of convex operators and respectively (ii) solving Markov moment problems. Recall also that, according to Chapter 1, solving abstract Markov moment problem leads to solving classical Markov moment problems, via additional results.

Remark 4.2. Given a vector space E, application of Theorem 4.2 to the very particular case when we endow E with the order relation defined by the equality-relation ($E_+ = \{0\}$), leads to the classical Hahn-Banach extension result (where positivity of T is not involved in any way).

An important particular case of Theorem 4.2 is that of sublinear operators as special convex operators Φ. In this respect, the following remarks are essential.

Remark 4.3. *Let E be a Banach space, F an order complete Banach space, $P \colon E \to F$ a continuous sublinear operator. Then for each $x_0 \in E$ and any $T \in \partial_{x_0} P$ we have $T(x_0) = P(x_0), T \in \partial P$. Consequently, $\partial P = \partial_0 P = \bigcup_{x_0 \in E} \partial_{x_0} P$.*

Indeed, if $T \in \partial_{x_0} P$, then, by definition, $P(x) - P(x_0) \geq T(x) - T(x_0)$ for all $x \in E$. Writing this for rx instead of $x, r \in [0, \infty)$, one obtains (via positively homogeneity):

$$r\big(P(x) - T(x)\big) \geq P(x_0) - T(x_0), x \in E$$

If $r = 0$ we infer that $T(x_0) \geq P(x_0)$. On the other hand, dividing by $r > 0$ and passing to the limit as $r \to \infty$ we obtain $P(x) \geq T(x), x \in E$. In particular, we have obtained $T(x_0) = P(x_0)$ and $T(x) \leq P(x), x \in E$. The conclusion follows. For notations used in this remark and detailed information see [31] or any othe reference concerning subdifferentails of convex operators.

Remark 4.4. *Let E, F be Banach lattices and , $P \colon E \to F$ a continuous sublinear operator. If $T \colon E \to F$ is a linear operator such that $T \leq P$ on E, then $\|T\| \leq \|P\|$ in the following two cases at least: 1) P is isotone; 2) P is symmetric ($P(x) = P(-x) \; \forall x \in E$).*

Indeed, in case 1), if P is isotone on E, then

$$T(x) \leq P(x) \leq P(|x|), -T(x) = T(-x) \leq P(-x) \leq P(|x|)$$

implies $|T(x)| \leq P(|x|), x \in E$, which further yields

$$\|T(x)\| \leq \|P(|x|)\| \leq \|P\| \|x\|, x \in E \implies \|T\| \leq \|P\|$$

In case 2), assuming that P is symmetric, we obtain

$$T(x) \le P(x), -T(x) = T(-x) \le P(-x) = P(x) \Longrightarrow$$

$$|T(x)| \le P(x) \Longrightarrow \|T(x)\| \le \|P(x)\|, x \in E \Longrightarrow \|T\| \le \|P\|$$

The preceding remark is still valid for convex symmetric operators $P: E \to F$ verifying

$$\sup_{\|x\| \le 1} \|P(x)\| \le M < \infty, P(0) = 0$$

In this case, if T is linear, dominated by P on the entire domain space E, then $\|T\| \le M$. Indeed, similar to the case 2) from above, we have: $\|T(x)\| \le \|P(x)\|, x \in E$. This leads to

$$\|T\| = \sup_{\|x\| \le 1} \|T(x)\| \le \sup_{\|x\| \le 1} \|P(x)\| \le M$$

Examples of such convex operators which are not sublinear can be constructed defining

$$P(x) = \|x\|^p y_1, x \in E, p \in (1, \infty),$$

where $y_1 \in F_+, y_1 \ne 0$. For the above operator, the constant M equals $\|y_1\|$. Other examples of such convex operators that are not sublinear can be obtained forming linear combinations with positive coefficients of operators emphasized by the last formula.

Remark 4.5. It seems that in the proof of Corollary 4.1 the continuity of P on E_+ is not used.

References

[1] N.I. Akhiezer, *The Classical Moment Problem and Some Related Questions in Analysis*, Oliver and Boyd, Edinburgh-London, 1965.

[2] C. Ambrozie, O. Olteanu, *A sandwich theorem, the moment problem, finite-simplicial sets and some inequalities*, Rev. Roumaine Math. Pures Appl., **49**, 3(2004), 189-210.

[3] V. Balan, A. Olteanu and O. Olteanu, *On Newton's method for convex operators with some applications*, Rev. Roumaine Math. Pures Appl., **51**, 3(2006), 277-290.

[4] C. Berg, J.P.R. Christensen and C.U. Jensen, *A remark on the multidimensional moment proble*,MathematischeAnnalen,243(1979),163-169.
https://doi.org/10.1007/BF01420423

[5] C. Berg and A.J. Durán, *The fixed point for a transformation of Hausdorff moment sequences and iteration of a rational function*, Math. Scand. 103 (2008), 11-39.
https://doi.org/10.7146/math.scand.a-15066

[6] C. Berg and M. Beygmohammadi, *On fixed point in the metric space of normalized Hausdorff moment sequences*, Rend. Circ. Mat. Palermo, serie II, Suppl. 82 (2010), 251-257.

[7] N. Boboc, Gh. Bucur, *Convex cones of continuous functions on compact spaces*, Academiei, Bucharest, 1976 (Romanian).

[8] H. Bonnel, J. Collonge, *Optimization over the Pareto outcome set associated with a convex bi-objective optimization problem: theoretical results, deterministic algorithm and application to the stochastic case*, Journal of Global Optimization, 62(3) (2015), 481-505.
https://doi.org/10.1007/s10898-014-0257-0

[9] H. Bonnel, C. Schneider, *Post Pareto Analysis and a New Algorithm for the Optimal Parameter Tuning of the Elastic Net*, Journal of Optimization Theory and Applications (2019)
https://doi.org/10.1007/s10957-019-01592-x

[10] G. Cassier, *Problèmes des moments sur un compact de $\mathbb{R}^n$ et décomposition des polynômes à plusieurs variables*, Journal of Functional Analysis, 58 (1984), 254-266.
https://doi.org/10.1016/0022-1236(84)90042-9

[11] G. Choquet, *Le problème des moments*, Séminaire d'Initiation à l'Analise, Institut H. Poincaré, Paris, 1962.

[12] I. Colojoară, *Elements of spectral theory*, Academiei, Bucharest, 1968 (Romanian).

[13] R. Cristescu, *Ordered Vector Spaces and Linear Operators*, Academiei, Bucharest, Romania, and Abacus Press, Tunbridge Wells, Kent, England, 1976.

[14] C. Drăgușin, *Min-max pour des critères multiples*, RAIRO Recherche Opérationnelle/OperationsResearch,**12**,2(1978),169-180.
https://doi.org/10.1051/ro/1978120201691

[15] B. Fuglede, *The multidimensional moment problem*, Expo. Math. I (1983), 47-65.

[16] L. Gosse and O. Runborg, *Existence, uniqueness, and a constructive solution algorithm for a class of finite Markov moment problems*, SIAM J. Appl. Math., 68, 6 (2008),1618-1640.
https://doi.org/10.1137/070692510

[17] E.K. Haviland, *On the momentum problem for distributions in more than one dimension*,Amer.J.Math.58(1936),164-168.
https://doi.org/10.2307/2371063

[18] R.B. Holmes, *Geometric Functional Analysis and its Applications*, Springer, 1975.
https://doi.org/10.1007/978-1-4684-9369-6

[19] C. Kleiber, J. Stoyanov, *Multivariate distributions and the moment problem*, JournalofMultivariateAnalysis,113(2013),7-18.
https://doi.org/10.1016/j.jmva.2011.06.001

[20] M. G. Krein and A. A. Nudelman, *Markov Moment Problem and Extremal Problems*, American Mathematical Society, Providence, RI, 1977.

[21] S.S. Kutateladze, *Convex operators*, Russian Math. Surveys **34** 1(1979), 181-214.

[22] L. Lemnete, *An operator-valued moment problem*, Proc. Amer. Math. Soc. 112, (1991), 1023-1028.
https://doi.org/10.1090/S0002-9939-1991-1059628-5

[23] L. Lemnete-Ninulescu and A. Zlătescu, *Some new aspects of the L-moment problem*, Rev. Roumaine Math. Pures Appl. **55**(3), (2010), 197-204.

[24] M. Marshall, *Positive Polynomials and Sums of Squares*, American Mathematical Society,2008.
https://doi.org/10.1090/surv/146

[25] M. Marshall, *Polynomials non-negative on a strip*, Proceedings of the American MathematicalSociety,**138**,(2010),1559-1567.
https://doi.org/10.1090/S0002-9939-09-10016-3

[26] J.M. Mihăilă, O. Olteanu, C. Udrişte, *Markov-type and operator-valued multidimensional moment problems, with some applications*, Rev. Roum Math Pures Appl., **52**, 4 (2007), 405-428.

[27] J.M. Mihăilă, O. Olteanu, C. Udrişte, *Markov - type moment problems for arbitrary compact and some noncompact Borel subsets of* $\mathbb{R}^n$, Rev. Roum Math Pures Appl. **52**, 5-6 (2007), 655-664.

[28] M.M. Newmann, *Uniform boundedness and closed graph theorems for convex operators*,MatematischeNachrichten,**120**,1(1985),113-125.
https://doi.org/10.1002/mana.19851200111

[29] C. Niculescu, N. Popa, *Elements of Theory of Banach Spaces*, Academiei, Bucharest, 1981 (Romanian).

[30] C.P. Niculescu, L..-E. Persson, *Convex Functions and Their Applications. A Contemporary Approach*, 2nd Ed., CMS Books in Mathematics Vol. **23**, Springer-Verlag, New York, 2018.

[31] C.P. Niculescu and O. Olteanu, *A note on the isotonic vector-valued convex functions*, Preprint, arXiv:2005.01088v1 [math.FA] 3 May 2020.

[32] D.T. Norris, *Optimal Solutions to the L_∞ Moment Problem with Lattice Bounds*, PhD Thesis. (2002), University of Colorado, Mathematics Department, 2003

[33] V.V. Olariu, C.O. Olteanu, *Orthogonality and Mathematical Physics*, LAMBERT Academic Publishing, Beau Bassin, 2018.

[34] O. Olteanu, *Convexité et prolongement d'opérateurs linéaires*, C. R. Acad. Sci. Paris, Série A, **286** (1978), 511-514.

[35] O. Olteanu, *Sur les fonctions convexes définies sur les ensembles convexes bornés de $\mathbb{R}^n$*, C..R. Acad. Sci. Paris, Série A, **290** (1980), 837-838.

[36] O. Olteanu, *Théorèmes de prolongement d'opérateurs linéaires*, Rev. Roum Math. Pures Appl., **28**, 10 (1983), 953-983.

[37] O. Olteanu, *Application de théorèmes de prolongement d'opérateurs linéaires au problème des moments e à une généralization d'un théorème de Mazur-Orlicz*, C. R. Acad Sci Paris, t. 313, Série I, (1991), 739-742.

[38] O. Olteanu, *A strong separation theorem in normed linear spaces*, Mathematica, (Cluj), **35(58)**, 1 (1993), 59-63.

[39] O. Olteanu, *Applications of a general sandwich theorem for operators to the moment problem*. Rev. Roumaine Math. Pures Appl., **41**, 7-8 (1996), 513-521.

[40] O. Olteanu, *New results on Markov moment problem*, International Journal of Analysis, Vol. 2013, Article ID 901318, pp. 1-17.
http://dx.doi.org/10.1155/2013/901318

[41] O. Olteanu, *Moment problems on bounded and unbounded domains*, International Journal of Analysis, Vol. 2013, Article ID 524264, pp. 1-7.
http://dx.doi.org/10.1155/2013/524264

[42] O. Olteanu, *Approximation and Markov moment problem on concrete spaces*, Rendiconti del Circolo Matematico di Palermo, **63** (2014), 161-172.
https://doi.org/10.1007/s12215-014-0149-7

[43] O. Olteanu, *Applications of Hahn-Banach principle to the moment problem*, Poincare Journal of Analysis and Applications, **1** (2015), 1-28.

[44] O. Olteanu, *Markov moment problems and related approximation*, Mathematical Reports, **17(67)**, 1 (2015), 107-117.

[45] O. Olteanu, *Invariant subspaces and invariant balls of bounded linear operators*, Bull. Math. Soc. Sci. Math. Roumanie, **59** (107) 3(2016), 261-271.

[46] O. Olteanu, J.M. Mihăilă, *Operator - valued Mazur - Orlicz and moment problems in spaces of analytic functions*, U. P. B. Sci. Bull., Series A, **79**, 1 (2017), 175-184.

[47] O. Olteanu, *Mazur-Orlicz theorem in concrete spaces and inverse problems related to the moment problem*, U.P.B. Sci. Bull. Series A, **79**, 3 (2017), 151-162.

[48] O. Olteanu, J. M. Mihăilă, *On Markov moment problem and Mazur - Orlicz theorem*, Open Access Library Journal, 4(2017), 1-10. https://doi.org/10.4236/oalib.1103950.

[49] O. Olteanu, J. M. Mihăilă, *Extension and decomposition of linear operators dominated by continuous increasing sublinear operators*, U.P.B. Sci. Bull. Series A, 80, 3 (2018), 133-144.

[50] O. Olteanu, J. M. Mihăilă, *Polynomial approximation on unbounded subsets and some applications*, Asian Journal of Science and Technology, 9, 10(2018), 8875-8890.

[51] O. Olteanu, J.M. Mihăilă, *A class of concave operators and related optimization*, U.P.B. Sci. Bulletin Series A, **81**, 3 (2019), 165-176.

[52] O. Olteanu, J.M. Mihăilă, *Markov moment problem in concrete spaces revisited*, MathLAB Journal, **5** (2020), 82-91.

[53] O. Olteanu, *New results on extension of linear operators and Markov moment problem*, MathLAB Journal, **5** (2020), 143-154.

[54] O. Olteanu, *New results in mathematical analysis*, LAMBERT Academic Publishing, Beau Bassin, 2020.

[55] O. Olteanu, *On Newton's method for convex functions and operators and its relationship with contraction principle*, MathLAB Journal (accepted).

[56] R.R. Phelps, *Lectures on Choquet's Theorem. Second Edition*, Springer-Verlag, Berlin Heidelberg, 2001.

[57] M. Putinar, *Positive polynomials on compact semi-algebraic sets*, Indiana University Mathematical Journal, 42, 3 (1993), 969-984. https://doi.org/10.1512/iumj.1993.42.42045

[58] W. Rudin, *Real and Complex Analysis. Third Edition*, McGraw-Hill Book Company International Edition, 1987.

[59] H.H. Schaefer, *Topological Vector Spaces. Third Printing Corrected*, Springer Verlag, 1971. https://doi.org/10.1007/978-1-4684-9928-5

[60] K. Schmüdgen, *The moment problem for compact semi-algebraic sets*, Mathematische Annalen,289(1991),203-206.

https://doi.org/10.1007/BF01446568

[61] K. Schmüdgen, *The Moment Problem*. Graduate Texts in Mathematics, Springer, 2017. https://doi.org/10.1007/978-3-319-64546-9.

[62] J. Stoyanov, G.D. Lin, *Hardy's condition in the moment problem for probability distributions*, Theory Probab. Appl. **57**, 4 (2013) (SIAM edition), 699-708. https://doi.org/10.1137/S0040585X9798631X

[63] M. Valadier, *Sous-Différentiablité de fonctions convexes à valeurs dans un espace vectoriel ordonné*. Math. Scand. 30, (1972), 65-74. https://doi.org/10.7146/math.scand.a-11064

[64] F.H. Vasilescu, *Spectral measures and moment problems*. In "Spectral Analysis and its Applications. Ion Colojoară Anniversary Volume", Theta, Bucharest, 2003, pp. 173-215.

[65] K. Yosida, *Functional Analysis. Sixth Edition*, Springer-Verlag, 1980.

[66] J. Zowe, *Linear Maps Majorized by a Sublinear Map*, ARCH. MATH. 26 (1975), 637-645.
https://doi.org/10.1007/BF01229793

[67] J. Zowe, *Sandwich Theorems for Convex Operators with Values in an Ordered Vector Space*, J.Math. Anal. Appl. 66, (1978), 282-296.
https://doi.org/10.1016/0022-247X(78)90232-9

www.ingramcontent.com/pod-product-compliance
Lightning Source LLC
Chambersburg PA
CBHW050552160726
48003CB00002B/868